REEDS INTRODUCTIONS

Principles of Earth Observation
For Marine Engineering Applications

REEDS INTRODUCTIONS

Essential Sensing and Telecommunications for Marine Engineering Applications

Physics Wave Concepts for Marine Engineering Applications

Principles of Earth Observation for Marine Engineering Applications

REEDS INTRODUCTIONS

Principles of Earth Observation
For Marine Engineering Applications

Christopher Lavers

REEDS
LONDON • OXFORD • NEW YORK • NEW DELHI • SYDNEY

REEDS
Bloomsbury Publishing Plc
50 Bedford Square, London, WC1B 3DP, UK

BLOOMSBURY, REEDS, and the Reeds logo are trademarks of Bloomsbury Publishing Plc

First published in Great Britain 2019

Copyright © Christopher Lavers, 2019

Christopher Lavers has asserted his right under the Copyright, Designs and Patents Act, 1988, to be identified as Author of this work

All rights reserved. No part of this publication may be reproduced or transmitted in any form or by any means, electronic or mechanical, including photocopying, recording, or any information storage or retrieval system, without prior permission in writing from the publishers

Bloomsbury Publishing Plc does not have any control over, or responsibility for, any third-party websites referred to or in this book. All internet addresses given in this book were correct at the time of going to press. The author and publisher regret any inconvenience caused if addresses have changed or sites have ceased to exist, but can accept no responsibility for any such changes

A catalogue record for this book is available from the British Library

Library of Congress Cataloguing-in-Publication data has been applied for.

ISBN: PB: 978-1-4729-4999-8; ePub: 978-1-4729-5000-0; ePDF: 978-1-4729-5001-7

2 4 6 8 10 9 7 5 3 1

Typeset in Myriad Pro by Newgen KnowledgeWorks (P) Ltd, Chennai, India
Printed and bound in Great Britain by CPI Group (UK) Ltd, Croydon CR0 4YY

To find out more about our authors and books visit www.bloomsbury.com
and sign up for our newsletters

Dedication

There are many whom I would like to thank in the preparation of this book, which has been both a challenge and a delight to write. I would like to express particular professional thanks to Dr Sam Lavender (Pixalytics) for use of imagery provided and helpful insights into several specific topic areas, to Dr Travis Mason (NOC Southampton) for ongoing collaborative aspects related to image processing of Earth imagery and access to valuable data sets, and lastly Dr Tim Absalom (Plymouth University), who has always been immensely helpful in sorting out my issues with ArcGIS (various versions). I would also like to thank the 'proofreading section' provided by my wife, Anne, BA (Hons) Geography, my daughter, Sara-Kate, and my son, Matthew; sometimes, in terrestrial Earth observation, it is indeed difficult to see the wood for the trees!

CONTENTS

Introduction 1

1 Monitoring the Earth environment: requirements, historical review and key principles 6

1.1 Requirements for Earth observation in the 21st century 6
1.2 A brief historical review of Earth observation 9
1.3 Classification of remote sensing systems 12
1.4 Electromagnetic radiation 12
1.5 Remote sensing application sensing bands 17
1.6 Some common remote sensing terms and units 20
1.7 Resolution issues 22
1.8 Attenuation and radiation propagation 22
1.9 Albedo 24
1.10 Spectral radiant flux, reflectance, absorbance and transmittance 24
1.11 Radiation emission 26
1.12 Hemispherical reflectance 29
1.13 A remote sensing system 30
1.14 The remote sensing process 31
Questions 32
References 33

2 Visible, near infrared and ultraviolet electromagnetic radiation interactions at the Earth's surface 36

2.1 The interaction of visible, NIR and ultraviolet with Earth's surface 36
2.2 Water properties 38
2.3 The interaction of visible light and NIR with water (the hydrosphere) 41
2.4 Underwater light attenuation 42
2.5 The interaction of ultraviolet with water (the hydrosphere) 47
2.6 Vegetation 48
2.7 Time dependent characteristics 51
2.8 Canopy geometry changes 53
2.9 The interaction of UV with vegetation 53
2.10 Vegetation, and normalised difference Vegetation Index 53
2.11 The interaction of visible and NIR with soil 55

2.12 The interaction of visible, NIR and ultraviolet radiation with rock
and minerals   58
2.13 The interaction of ultraviolet with rocks and minerals   58
2.14 The interaction of visible and NIR radiation with snow and ice (the
cryosphere)   58
2.15 The interaction of ultraviolet with snow and ice (the cryosphere)   59
Questions   60
References   61

3 Thermal sensors   64

Introduction   64
3.1 Thermal radiation and its interactions with the Earth's surface   64
3.2 Emissivity   65
3.3 Spatial variability   70
3.4 Principal wavebands   70
3.5 Kinetic temperature   70
3.6 Thermal crossover   72
3.7 Heating rate   72
3.8 Interaction of thermal infrared wavelengths of electromagnetic radiation
with water (the hydrosphere)   72
3.9 Interaction of thermal infrared wavelengths of electromagnetic radiation
with vegetation and chlorophyll   73
3.10 Interaction of thermal infrared wavelengths of electromagnetic
radiation with snow and ice   74
3.11 Interaction of thermal infrared wavelengths of electromagnetic
radiation with soil   75
3.12 Interaction of thermal infrared wavelengths of electromagnetic
radiation with rocks and minerals   76
3.13 Satellite thermal IR systems   76
3.14 Thermal IR spectra   79
3.15 Non-imaging systems   79
3.16 Far infrared thermal imaging sensors   82
3.17 Detectors for thermal infrared radiation   83
3.18 Advanced cooled FPAs   85
3.19 Emerging Uncooled FPA (UFPA)   86
Questions   86
References   88

4 Microwave sensors — 90

4.1 Problems with visible imagery	90
4.2 Passive microwave sensors	90
4.3 Active microwave sensors	91
4.4 The echo ranging principle	92
4.5 Radar parameters	93
4.6 Doppler radar	95
4.7 Radar antennas	95
4.8 Phased arrays	96
4.9 Radar imaging	100
4.10 Inverse Synthetic Aperture Radar (ISAR)	109
4.11 Sensor system types	109
4.12 Interaction of microwaves with different surfaces	111
Questions	114
References	115

5 Atmospheric interactions with electromagnetic radiation — 116

5.1 Radiation from the sun and the solar radiation spectrum	116
5.2 The atmospheric absorption spectrum	117
5.3 Atmospheric transmission	118
5.4 Radiation from Earth	119
5.5 Atmospheric composition	120
5.6 Atmospheric and ionospheric turbulence	123
5.7 Cloud, rain and snow	123
5.8 Radiation propagation	123
5.9 Absorbance and transmittance	125
5.10 Ocean attenuation	125
5.11 The remote sensing inverse problem	128
Questions	129
References	130

6 Hydrosphere and cryosphere applications — 131

6.1 Water resource applications: the hydrosphere and the cryosphere	131
6.2 Water pollution detection	132
6.3 Lake eutrophication	133
6.4 Ice shelves *visible*	134

6.5 Water security issues	134
6.6 Ocean colour *visible*	136
6.7 Ocean wind *microwave*	137
6.8 Rivers	138
6.9 Wetland mapping	139
6.10 Surveillance maritime applications	140
6.11 Oil spillages	147
6.12 Sea and ice radar interferometry	149
Questions	150
References	151

7 Land resource applications 153

7.1 Land resource applications	153
7.2 Land cover	153
7.3 Rocks	154
7.4 Geological mapping	154
7.5 Soil mapping and evaluation	158
7.6 Soil salinity	159
7.7 Land use/cover mapping classification	160
7.8 Vegetation cover	162
7.9 Forest applications	163
7.10 Archaeology	164
7.11 Land glaciers *visible*	166
7.12 Urban and regional planning applications	167
7.13 Land surveillance	169
7.14 Disaster monitoring	170
Questions	171
References	172

8 Atmospheric applications 175

8.1 Atmospheric remote sensing applications	175
8.2 Measurement geometries	178
8.3 Atmospheric layer sensing	179
8.4 Satellite validation principles	190
8.5 Available products	190
Questions	191
References	192

9 Satellite platforms for remote sensing   194

9.1 An early history of non-terrestrial platforms	194
9.2 Rocketry	195
9.3 What is a satellite?	196
9.4 Satellite physics basics	197
9.5 Types of satellite	201
9.6 Comparison of polar orbiting and geostationary satellites	202
9.7 Weather sensing satellites	203
9.8 Earth observation satellites	204
9.9 High-resolution satellite missions	208
9.10 Small satellite missions and nanosats	208
9.11 Other notable Earth observation satellites	209
Questions	212
Reference	214

10 Introduction to satellite image processing and other imagery sources   215

10.1 Introduction to image processing	215
10.2 Pre-processing	215
10.3 Image enhancement	217
10.4 Image transformations	223
10.5 Image Interpretation	227
10.6 Change detection	228
10.7 Image classification and analysis	229
10.8 Other imagery sources	230
10.9 Web-based satellite image sources	231
Questions	233
References	234
Appendix 1: Numerical solutions	236
Glossary	242
Index	247

Introduction

This book introduces maritime and terrestrial Earth observation, including the *littoral* environment, to engineers, physicists and those interested in this subject. The terms *Earth observation* or, more commonly, *remote sensing* have been used interchangeably over recent decades, but require qualification. Although remote sensing can be viewed in wonderful coffee-table books [11–13], we will view the subject from a scientific perspective. This volume examines in greater detail aspects not widely covered in other topical books, partly due to the rapid advances and changes currently taking place in this field, such as atmospheric monitoring and digital signal processing of Earth observation images. The atmosphere has perhaps for too long been seen as an irritating medium 'to be corrected for' rather than a medium to be appreciated in its own right. The advent of digital signal processing of satellite and aerial platform images has drawn from innovations in medical digital image processing, and fruitful overlaps are expected in these areas over the coming decades. Development of maturing drone technology as a robust, digitally secure platform will provide opportunities in dangerous and awkward remote sensing applications, and in areas where remote land and maritime use is required for both civilian and military users.

Remote sensing is regarded as acquisition of information about the Earth without making physical contact with it, and it is generally assumed to be through use of satellite-based sensors. This isn't always the case, as aircraft-based sensors and, increasingly, Unmanned Aerial Vehicles (UAVs) bear testimony. Earth observation largely refers to satellite-based remote sensing, and seeks to highlight the *focus* of the sensing. Earth Observation (EO) gathers information about Earth's physical, biological and chemical systems via remote sensing methods, Earth surveying techniques, including collection, with analysis, and data presentation using Geographical Information System (GIS) methods. Ambiguities in the term 'Earth observation' are avoided by appeal to the specific platform involved, such as UAV remote sensing or satellite remote sensing.

What is remote sensing?

The term 'remote sensing' was coined in the 1950s by Ms Evelyn Pruitt, a geographer in the US Office of Naval Research, to differentiate between the new high-altitude aircraft images and, later, new satellite data sources as distinct from standard airborne photography [14]. To understand what remote sensing is, it is helpful to consider definitions put forward by experts in the field.

Principles of Earth Observation

> **Definitions of remote sensing**
>
> 'A group of techniques for collecting images or other forms of data about an object from measurements made at a distance from the object, and the processing and analysis of the data.' (CCRS-RESORS)
>
> 'Earth Observation (EO) is about observing and monitoring planet Earth, primarily by a methodology called remote sensing, which is essentially collection of data by sensors on airborne (such as aircraft and drones) and satellite platforms, which is processed by computer systems to provide information and images about a geographical area that can range from the global to local scale. In addition, "Earth Observations" encompasses the collection and analysis of a much broader range of measurement techniques. At the local scale the sensors could also be held in the hand or placed on moving and fixed platforms, e.g. there are a number of tall vertical platforms topped with sensors in the Amazon and marine measurements can be taken from vessels, buoys and floats alongside autonomous vehicles. The information gathered can be related to physical, chemical and biological systems and used for applications that relate to both the natural world and anthropogenic activities. Observations are often combined with empirical models that allow us to look beyond what's measurable using our current system understanding.' Dr Sam Lavender, CEO, Pixalytics
>
> 'Remote Sensing: the science and art of obtaining useful information about an object, area or phenomenon through analysis of data acquired by a device that is not in contact with the object, area or phenomenon under investigation.' [I5]
>
> 'Remote sensing may be broadly defined as the collection of information about an object without being in physical contact with the object. Aircraft and satellites are the common platforms from which remote sensing observations are made. The term "remote sensing" is restricted to methods that employ electromagnetic energy as the means of detecting and measuring target characteristics.' [I6]

In the broadest sense, remote sensing is the measurement and analysis of properties of Earth's atmosphere, land, hydrosphere or cryosphere using a recording device not in physical contact with the surface under study. The sensor is used remotely from air or space, gathering information through measurement of absorbed, transmitted or reflected electromagnetic radiation. The *technique* employs devices such as radar, laser, digital camera or radio frequency sounders. It is the *practice* of collecting data in wavelengths routinely from the ultraviolet to the microwave region and is the practical development of airborne photography through improvements in sensing technology, data processing, and advances in aviation and satellite platforms.

This book provides an all-round basic knowledge of Earth observation, including fundamental principles and current technological developments, with application to maritime and terrestrial real-world problems. It will help those interested to gain a significant understanding of the issues involved, aid those managing remotely sensed data, particularly in the maritime environment, and develop the reader's capacity for original thinking.

It has been said that Earth observation is like a reading process, with the words on the page viewed 'remotely' – letters contrasted against their background with the brain processing light and dark variations to provide understanding! In a similar way, we use many man-made sensors to remotely collect and analyse information about Earth. Data comes in many forms: electromagnetic, acoustic, gravimetric and a range of forces. This book focuses primarily on electromagnetic wave sensor use, and confines itself to measurement of 'surface targets', often in isolated terrestrial and maritime locations. Although there are several mostly electromagnetic sensing systems in common hospital circulation – X-rays, CAT scans, MRI, thermal and laser imaging and endoscopy – these applications are omitted, as is subsurface acoustic sonar, except where they directly relate to surface Earth observation.

The two basic processes involved in ElectroMagnetic (EM) remote sensing of Earth resources are **Data Acquisition** and **Data Analysis**, leading to **Data Utilisation**, and these are interrelated (figure I1). Typical elements of Data Acquisition include: (a) an appropriate energy source, (b) energy propagation through the atmosphere from Earth's surface, and frequently as echoes from active sensors, (c) interactions with Earth surface features, (d) the response of aircraft or space-based sensors and (e) generation of images or numerical arrays.

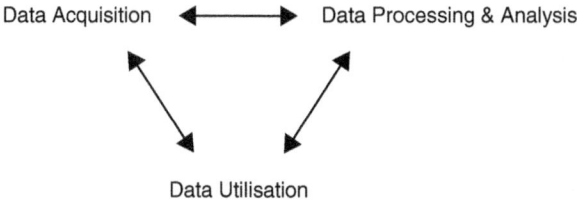

Figure I.1 *Data Acquisition and Data Analysis.*

Sensors record Earth features in reflected or emitted electromagnetic energy. Data Analysis involves viewing, analysis and interpretation of imagery or analysis of data arrays. With the aid of reference data, an analyst, or increasingly machine-based software, extracts critical information about the type, extent, location and resource condition from multiple data types. Information is represented in forms readily understandable by humans, a combination of maps, tables and written reports. Typical data products produce weather maps, pollution spillage and forest fire location, land use, crop area statistics or mineral locations. This information is presented to the end user (the customer), who applies this to inform their decision-making processes and is termed Data Utilisation, the 'application' layer of the process.

Remote sensing is addressing many varied key challenges in the oceans, atmosphere, cryosphere and land surfaces, including quantifying ocean dynamics and understanding the physical and biochemical air-sea land-sea interactions, aided by modelling. One of the key atmospheric challenges is quantifying and better modelling natural and anthropogenic-induced climate change and understanding the life cycle of aerosols and their interactions with clouds from observation of pollutant gases. In the cryosphere, it is important to define sea ice and land ice distributions, with the dynamic feedbacks to ocean-atmosphere, and the impact of snow ice on the global water cycle, permafrost changes and the land ecosystems, exchange of water and carbon from sea-land, air-land interactions.

There already exists a range of books dedicated to in-depth study, but several are worth noting for the serious student [I9–I13]. It is not my intent to repeat their work but to 'stand on the shoulders of giants' and focus on introductory aspects relevant to understanding of Earth observation principles for maritime and terrestrial applications that relate to engineers, scientists and physicists interested in gathering information about the Earth's land and maritime surfaces. I have covered much of the key background understanding of electromagnetic waves and wave principles elsewhere [I7–I8] but essential principles are covered in this volume.

REFERENCES

[I1] *Earthwatch, A Survey of the World from Space*, Charles Sheffield (Sidgwick & Jackson, London, 1981, ISBN 0283987375).

[I2] *The Atlas of Natural Wonders*, Rupert O Matthews (Ebury Press, London, 1988, ISBN 0852237758).

[I3] *The Earth from the Air 366 Days*, Yann Arthus-Bertrand (Thames & Hudson, London, 2003, ISBN 9780500542781).

[I4] earthobservatory.nasa.gov/Features/RemoteSensing/

[I5] *Remote Sensing and Image Interpretation*, Thomas Lillesand, Ralph W Kiefer and Jonathan Chipman (John Wiley & Sons, Inc., New Jersey, 2015, ISBN 1118343289).

[I6] *Remote Sensing: Principles and Interpretation*, 3rd Edition, Floyd F Sabins (W.H. Freeman and Company, San Francisco, 1996, ISBN 9780716724421).

[I7] *Physics Wave Concepts for Marine Engineering Applications*, Christopher Lavers and Sara-Kate Lavers (Bloomsbury Publishing, London, 2017, ISBN 1472922151).

[I8] *Essential Sensing and Telecommunications for Marine Engineering Applications*, Christopher Lavers (Bloomsbury Publishing, London, 2017, ISBN 1472922182).

[I9] *Introduction to Remote Sensing*, 2nd Edition, Arthur P Cracknell and Ladson Hayes (CRC Press, Taylor & Francis Group, London, 2007, ISBN 9780415335799).

[I10] *Introduction to Remote Sensing*, 3rd Edition, James B Campbell (Taylor & Francis Group, London, 2002, ISBN 0415282942).

[I11] *Physical Principles of Remote Sensing*, 3rd Edition, WG Rees (Cambridge University Press, 2012, ISBN 9780521181167).

[I12] *Introduction to Environmental Remote Sensing*, 2nd Edition, EC Barrett and LF Curtis (Routledge, London, 1999, ISBN 0748740066).

[I13] *Principles of Remote Sensing*, Paul J Curran (Longman Group Limited, London, 1985, ISBN 0582300975).

1
Monitoring the Earth Environment: Requirements, Historical Review and Key Principles

> *'We regret we are unable to give you the weather. We rely on weather reports from the airport, which is closed because of the weather. Whether we are able to give you the weather tomorrow depends on the weather.'* Arab News, Jeddah, January 1979

1.1 Requirements for Earth observation in the 21st century

Earth observation sensors and techniques have significantly improved our ability to gather information about the entire Earth's environment, but why bother? An adequate answer may be broken down into several parts. It exists to:

1) improve our database or inventory of Earth's natural resources and manage them efficiently.

2) regularly monitor changes in local, national and international environments.

3) avoid short-term economic gain through ill-advised and often illegal international trade, and quantify endangered natural resources or species, such as fish stocks, the giant panda, mountain gorilla, black rhino etc.

4) provide remotely sensed imagery and data for various applications, such as Maritime Pollution (MarPol) monitoring.

In each case, more detailed information is needed than is obtained from ground-level surveys alone.

The world's natural resources are finite and rapidly depleted by an expanding global population. Consider 'Earth Overshoot Day' 2 August 2017 (my birthday); it was estimated that on this day, we had already consumed our annual quota of renewable resources – that is, we had used more from nature than our planet renews in the year [1.1]. Earth's resources and distribution are of concern to the Third World, where hundreds of millions suffer from malnutrition; the poorest have a life expectancy decades below the US average, although there have been improvements in under-five mortality rates, which have fallen since 1990 from 93.4 deaths per 1,000 live births (1990) to 40.8 (2016) [1.2]. Even in the world's developed countries, there are often availability instabilities in some 'basic' commodities, such as fruit, vegetables [1.3] and wheat [1.4]. Increasing shortages may run into tens of millions of tonnes – a large figure with uncertain range. Timely global information, particularly for annual crops, provides more accurate and reliable estimates and helps prevent avoidable shortages.

Environmental damage or deterioration is caused by nature as well as man – for example, hurricanes and earthquakes. Man-made pollution of land, sea, cryosphere and air is monitored regularly by national and local authorities with ground-based equipment. However, pollution effects and international resources frequently cross international boundaries. This is where periodic remote sensing from Earth-orbiting satellites can provide reliable and near real-time monitoring methods. Environmental pollution results from natural disasters such as flooding and hurricanes. Because of the unpredictability of these events, global coverage provided by satellites is beneficial in locating and assessing them. More detailed coverage, required by rescue forces, is acquired by remote sensing equipment on aircraft or UAVs, such as proposed Search and Rescue UAVs for Coastguard Agency surveillance [1.5].

Key remote sensing advantages are that the speed data is acquired from large and inaccessible areas that may be investigated with/without permission, such as Zimbabwe and Sudan [1.6–1.9]. An extensive summary of remote sensing applications to environmental issues is given elsewhere [1.10]. The full range of remote sensing applications is diverse but a few key areas are relevant to the reader:

- ***Meteorology***: For weather analysis, modelling and forecasting – for example, atmospheric temperature, pressure, water vapour and wind velocity [1.11].

- ***Oceanography***: Particularly Sea Surface Temperature (SST), ocean circulation current patterns and speed, wave energy, water quality and salinity assessments [1.11]. Synthetic Aperture Radar (SAR), known as imaging radar, provides routine shipping traffic monitoring, oil slick detection and sea ice monitoring for ice-breakers.

- ***Glaciology, cryosphere, geology***: Glaciology studies of ice and natural phenomena such as ice sheet distribution and motion, as in Greenland, Iceland and Antarctica. The cryosphere is the frozen part of the Earth system including frozen parts of the ocean, such as the waters surrounding the Arctic and Antarctica [1.12]. Geology is interested in rock type, geological faults and location, measuring and observing tectonic motion, and geomorphology.

- ***Topography***: Acquiring accurate elevation data to quantify natural and artificial physical features, and ***Cartography***: The practice of making maps that model Earth so spatial information can be communicated effectively, such as for planning or to provide accurate land use inventory, monitoring changes and resources [1.13].

- ***Agriculture***: The science or practice of farming, including soil cultivation for crops and rearing animals to provide food, wool and other products evaluated in terms of vegetation land cover, health, soil types, water content, forecast yields, burn scars, illicit crop detection etc. ***Forestry***: The science of planting, managing, detecting illegal deforestation and harvesting forest products. ***Botany***: The study of the physiology, structure, genetics, ecology, distribution, classification and economic importance of plants [1.14].

- ***Hydrology***: Water resources such as spring snow meltwater – for example, the Great Lakes and the Baltic. The study of the movement, distribution and quality of water, including water cycle, resources and sustainability [1.10]. Imaging radar provides soil moisture and wetness maps and tropical storm rainfall rates.

- ***Disaster control, natural hazards and emergency management***: The organisation and management of resources and responsibilities for dealing with humanitarian emergencies to mitigate against hazards and disasters, such as sand and dust storm warnings, avalanches, landslides, floodwater, disease, pollution etc [1.15]. This work is often coordinated by transnational organisations such as UNOSAT delivering imaging analysis and satellite solutions through its technology arm, UNITAR [1.16].

- **Military and security space-based applications**: These are varied [1.17], including reconnaissance and the development of the first global high-precision navigation system Navstar GPS [1.18]. It is important for the military to detect and identify military aircraft, radar and missile emplacements. Security aspects include monitoring airports, weapon storage and airfields, as well as civilian monitoring of critical infrastructure: oil, gas, nuclear, electrical plants, railways, etc.

Space-based remote sensing has traditionally been expensive. However, a cost-benefit analysis demonstrates its financial effectiveness, and much remote sensing activity is justified. The development costs of a satellite and putting it into Low Earth Orbit (LEO) can be hundreds of millions of dollars. RADARSAT-1 was estimated to cost, excluding launch, C$620 million [1.19] but new nano-satellite development will reduce this by several orders of magnitude [1.20–1.21]. Crop-derived data from satellites can be worth £100 million per year. The Allen study estimated Australia's agriculture could benefit by A$1,005–1,357 million by 2030 if a nationwide Global Navigation Satellite System (GNSS) network were established [1.22]. Satellite remote sensing data, particularly meteorological satellites for disaster warning, are estimated to have saved tens of thousands of lives, with NASA's CYclone GNSS (CYGNSS) planned [1.23].

1.2 A brief historical review of Earth observation

Remote sensing is traced back to the 4th century BC with Aristotle's *camera obscura* but has accelerated over recent centuries (table 1.1). Optical theoretical developments began in the 17th century, although glass lenses were known earlier, but practical remote sensing had to await advances in photography pioneered by Thomas Wedgwood, Louis Daguerre and Henry Fox Talbot in the 18–19th centuries. During the 19th century, radiation beyond the visible was discovered by Herschel (infrared), Ritter (ultraviolet) and Hertz (radio waves). In 1863, Maxwell developed electromagnetic theory, upon which understanding of these waves depended [1.24].

Airborne photography followed fast upon the heels of photography. The Montgolfier brothers launched the first hot-air balloon in 1783, and in 1858 Gaspard-Félix Tournachon took the first aerial photograph from a balloon. The earliest surviving photograph taken from a balloon was of Petit Bicêtre near Paris in 1859. Kites were soon used, and in 1890 Arthur Batut published a textbook on aerial photography. Balloons were established in warfare throughout the 18th and 19th centuries, with observation balloons in use during the Austrian siege of Venice (1849), the American Civil War (1861–65), the Napoleonic Wars (1799–1815), the Franco-Prussian War (1870–71) and the Russo-Japanese War (1904–05).

The next step was successful demonstration of manned heavier-than-air flight, the aeroplane, by the Wright brothers, Orville and Wilbur, on 17 December 1903 at Kitty Hawk, with aerial photographs from aircraft recorded from 1909, after the first plane launch from a ship from the USS *Pennsylvania* flight deck. Airborne photography was used on a regular basis in the First World War (1914–18) for military reconnaissance, and a close connection between environmental and military remote sensing persists today. During the interwar period, civilian aerial photography developed, notably in relation to cartography, agriculture and forestry. Cameras improved, as did aircraft, alongside stereo photographic mapping. Work was undertaken by John Logie Baird, the inventor of television, into airborne scanning systems for transmission of images to ground.

The Second World War (1939–45) brought developments including photographic reconnaissance and microwave radar. The planned German invasion of Britain was halted in part by German photographic reconnaissance of dense shipping concentrations along the English Channel and by British radar development. Night bombers used the first prototype imaging radar, presenting operators with a terrain map, the earliest form of today's Side-Looking Airborne Radar (SLAR) and Synthetic Aperture Radar (SAR) systems.

Year	Major milestones in Earth Observation development
4th century BC	*Camera obscura*, Aristotle
1800	Discovery of infrared, Sir William Herschel
1839	Start of photography, Louis Daguerre
1847	Infrared spectra shown to have the properties of visible light, AHL Fizeau and JBL Foucault
1858	Balloon photography, Gaspard-Félix Tournachon ('Nadar')
1867	A dynamical theory of electromagnetic waves, developed by James Clerk Maxwell
1909	Photography from aeroplane, Orville Wright

1914–18	First World War: aerial reconnaissance and surveillance
1920–30	Civilian development of aerial photography and photogrammetry
1926	First moving TV pictures, John Logie Baird
1930–40	Development of radar in UK, US and Germany
1939–45	Second World War: application of infrared film for camouflage detection
1950s	First use of the term 'remote sensing' by E Pruitt Plant disease detection using IR film, Robert Colwell
1960	TIROS weather satellite
1968	First hyperspectral image, Apollo 9
1972	Launch of Landsat 1, the first multispectral scanner
1970–80	Digital image processing and analysis development
1982	Launch of Landsat 4
1986	Launch of SPOT 1 satellite (CNES, France)
1998	Launch of SPOT 4
1999	Launch of IKONOS
2000	First hyperspectral satellite remote sensing, Hyperion (220 spectral bands)
2001	QuickBird metric high-resolution imagery
2013	Doves: flocks of satellites
2018	Earth-i: first commercial high-resolution satellite video constellation

Table 1.1: *Major milestones in remote sensing.*

By the 1960s, false-colour infrared film for 'camouflage detection', originally for the military, found application in vegetation through Robert Colwell's work of 1959. Throughout the 1960s sensors were placed in space, originally part of the 'Moon Race', and the same techniques applied to Earth observation were realised. The first multispectral space imagery was obtained from Apollo 6. Since then, radar has been placed in space, while the computer revolution introduced digital image processing. This is regarded as the beginning of modern remote sensing, in which air and space observations became routine events.

Following the Second World War, the military made huge advances. Military satellites revealed Soviet rocket bases after spy planes identified them in Cuba (1962). Meteorological (Met) satellites were launched, providing valuable data. There is no doubt the 'Space Race' served as a critical turning point; after this date, there was an explosion in the type and volume of Earth observation data available, requiring digital software-driven analysis and raising the first questions about renewable resources, environmental protection and how to store and retrieve large 'meta-data' sets. Landsat provided the first systematic, reproducible observations over large areas with seven spectral bands in sufficient detail for many potential applications.

1.3 Classification of remote sensing systems

Remote sensing systems are classified in many ways, with distinctions drawn between **Active** and **Passive** systems, **Imaging** and **Non-Imaging**. We also distinguish sensors by the wavelengths they detect.

Active systems illuminate targets from their own source, generally of moderate to high power level due to the considerable stand-off distances involved. Passive systems use other energy sources, mostly the sun. The choice between active or passive is determined by several factors. A passive system is inappropriate at wavelengths where insignificant radiation levels occur naturally, or if required for night-time operation or conditions of limited or degraded visibility, such as fog or cloud. An active system may be technically infeasible if the power required to obtain an echo is too large, as this requires a system too heavy or large for the platform in question.

Illuminating radiation sources can be tailored to particular tasks, such as observing Doppler shifts in reflected radiation to calculate the relative velocity of approaching vessels. Military requirements vary greatly; generally, passive sensors are chosen, as they don't compromise the observation platform. However, parameters such as range can *only* be found with active transmission. Regardless of whether military systems are active or passive, they prefer to operate 'under cover of darkness'.

All imaging systems measure radiation intensity as a function of spatial position, so a two-dimensional picture representation of intensity is required. Non-imaging systems do not measure radiation intensity as a function of position. One dimension is sufficient for image production, as platform motion, if supported in the perpendicular direction, achieves the required two-dimensional criterion. Figure 1.1 illustrates a simplistic division of remote sensing systems into active or passive, imaging or non-imaging techniques.

1.4 Electromagnetic radiation

Most Earth observation sensors use ElectroMagnetic radiation (EM) as the source. This book is primarily concerned with EM sensors. The EM source used depends on the sensor. The EM spectrum for remote sensing is the wavelength range from nanometres to metres, travelling at light speed, and propagating through the vacuum of space, Earth's atmosphere, and underwater. Electromagnetic radiation is viewed in spectral regions, figure 1.2 (see plate section), listed in table 1.2.

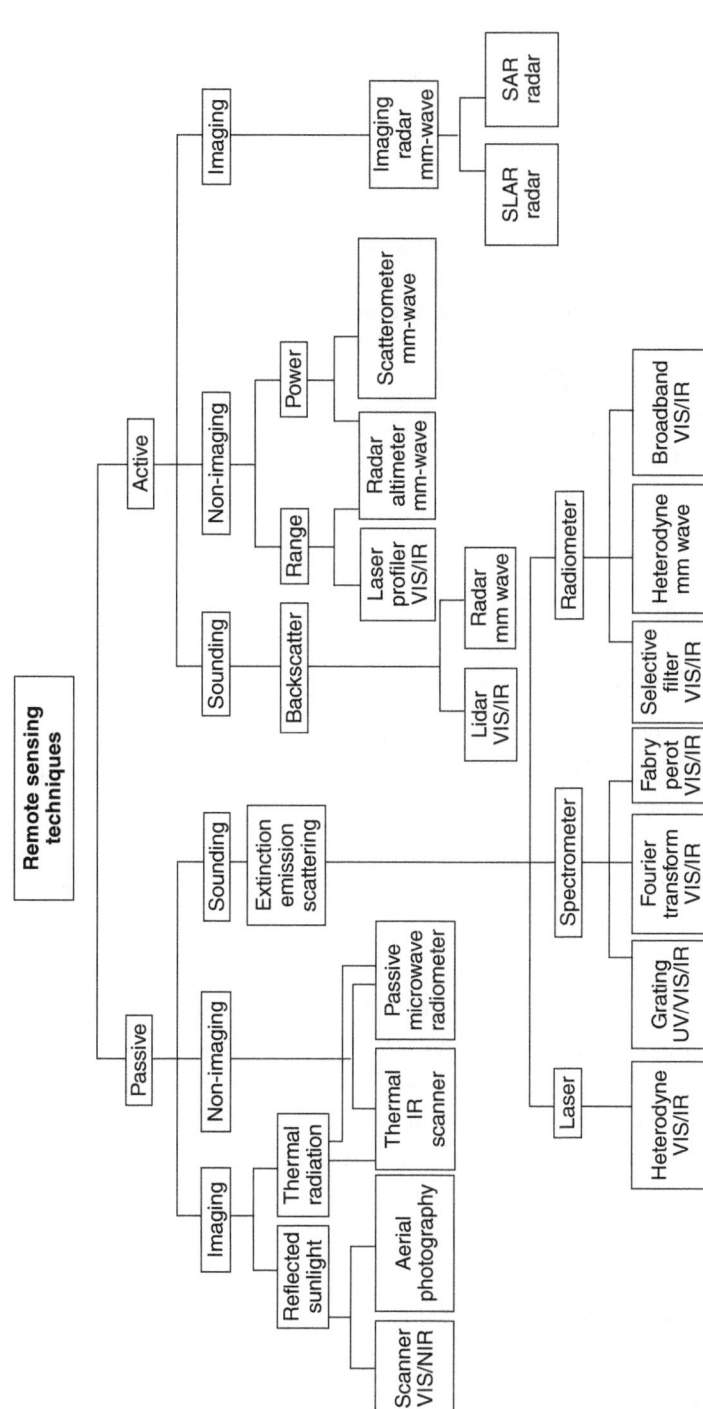

Figure 1.1: *Remote sensing techniques.*

Atmospheric transmission varies markedly with wavelength. A brief introduction to wavebands is given, but further details are found elsewhere [1.25].

Spectral region	Wavelength	Comment
Gamma ray	< 0.03nm	Completely absorbed by the upper atmosphere.
X-ray	0.03–30nm	Completely absorbed by the upper atmosphere.
Ultraviolet (UV)	0.2–0.4µm	Wavelengths below 0.28 microns are absorbed significantly by ozone in the upper atmosphere.
Visible	0.4–0.7µm	Images are taken with digital camera and photodetectors.
Reflected infrared	0.7–3.0µm	Wavelength dependent interaction with the Earth surfaces.
Thermal infrared	3.0–15µm	Heat images at these wavelengths are acquired with digital scanners and **not** with film.
Microwave radar	0.1–100cm	Active microwave remote sensing at various wavelength bands.
Radio	> 100cm	Long wave radio sounding measurements.

Table 1.2: *EM spectral regions used in remote sensing applications.*

1.4.1 Gamma rays < 0.03nm

These are used for treatment of diseases and examination of solid objects, and are produced through nuclear reactions. Paul Villard, a French physicist, is credited with their discovery in 1900.

1.4.2 X-rays 0.03–30nm

X-rays are lower in energy than gamma rays, sometimes called Röntgen rays after the German scientist Wilhelm Röntgen, who discovered them in 1895. X-rays pass through considerable thicknesses of solid substances such as brick or metal, which stops ordinary light. X-rays also pass easily through the human body, with denser bones absorbing more radiation than softer tissue. Modern medical X-ray machines convert transmitted radiation into digital images of the interior of the human body. Dental radiographers also use X-rays to examine teeth. The space-based SOlar and Heliospheric Observatory (SOHO) can obtain detailed X-ray images of the sun, besides other spectral bands.

1.4.3 Ultraviolet radiation 0.2–0.4µm

Ultraviolet (UV) radiation is shorter than visible wavelengths, and present in solar radiation. Ultraviolet is responsible for human sunburns, but is used by the body to produce vitamin D. Fortunately, high solar UV levels are absorbed by ozone (O_3) in the upper atmosphere, but high exposure levels are encountered at high altitude. Ultraviolet can detect certain substances through fluorescence (discussed in Chapter 2).

1.4.4 Visible 0.4–0.7μm

Radiation between 0.4 and 0.7 microns (or μm), that is, millionths of a metre, is where the human eye images and distinguishes many colours with precise discrimination, discussed further in Chapter 2.

1.4.5 Infrared 0.7–15.0μm

Infrared waves are longer in wavelength than visible red. Infrared is invisible to the human eye, but produces the sensation of heat on skin. Infrared was discovered in 1800 by astronomer Sir William Herschel when testing solar filters to observe sun spots.

1.4.6 Reflected near infrared 0.7–3.0μm

Short Near InfraRed (NIR) is used by Image Intensifiers (I^2) or 'night vision devices' for many applications and may be combined with low light level Closed Circuit TeleVision (CCTV). Although invisible to humans, devices detect invisible NIR from starlight, and amplify low level light present at night, from moonlight or urban lighting. NIR penetrates cloud and fog better than visible light (less scattering) but cloud, fog, sandstorm or smoke will reduce vision capabilities.

A typical intensifier has a green phosphor display as the human eye is more sensitive to green. Image intensifiers respond to visible and NIR and are displayed together as green unless filters are introduced. Daytime NIR image intensifiers can achieve longer range imaging than visible cameras, since NIR has reduced scattering and enhanced transmission through haze compared to visible light.

1.4.7 Thermal infrared 3.0–15μm

The Middle InfraRed (MIR), 3–6μm, is commonly associated with hot civilian sensing, such as erupting volcanoes [1.26], or military operation of heat-seeking missiles, which passively seek heat emissions from ships, such as a hot funnel outlined against cold seas and skies [1.27].

Far InfraRed (FIR) thermal cameras, operating at 6–15μm, cover the range associated with human and wildlife emissions. Search and Rescue (SAR) use thermal imaging extensively, providing identification at night in the absence of light or in mist and fog conditions [1.28]. Radar may *detect* a vessel, but a Thermal Imaging Camera (TIC) provides a vessel *picture*, useful for surveillance. However, these pictures are not the same as high-resolution visible or NIR, but interpret pictures of emitted and reflected longer wavelengths. TIC users require training to interpret imagery correctly, but it provides a useful maritime tool for SAR operations, surveillance, sea safety and vessel identification, among other applications: predictive maintenance, engine monitoring and Earth observation [1.28–1.29]. Modern commercial TICs are tools for properly trained maritime firefighters, and help contain fires successfully. In the marine environment, a TIC is an essential device today; unlike image intensifiers, it sees through thick black smoke. These systems are discussed elsewhere [1.30]. Cameras may be radiometrically calibrated, providing accurate temperatures, and are valuable in remote wildlife observation [1.31–1.32].

1.4.8 Microwaves 0.1–100cm

'Microwaves' is a description used for wavelengths between 1mm and 100cm, including UHF and SHF bands. Microwaves are used in household items today and in radar systems on air, sea, space and land platforms. Long-range systems operate below 4GHz, where transmission loss is lowest. Short-range systems operate up to 20GHz. Microwaves are generally 'small' compared with radio waves. Beginning at around 40GHz, the atmosphere becomes less transparent to microwaves due to lower frequency absorption from atmospheric water vapour and at higher frequencies from atmospheric oxygen. Above 100GHz, radiation absorption by Earth's atmosphere is so great it is effectively opaque, until the atmosphere becomes transparent again in the infrared and optical (figure 1.3).

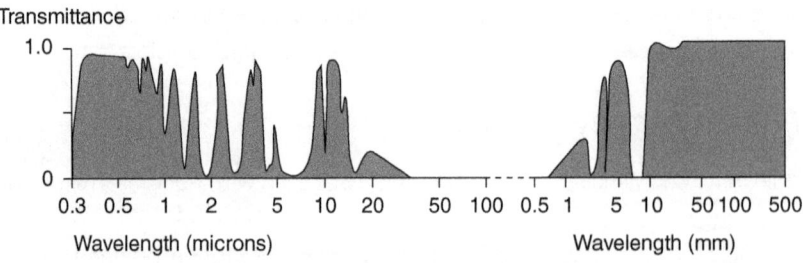

Figure 1.3: *Atmospheric transmittance.*

Maritime users extensively use RADAR (Radio Aid for Detection and Ranging), an active method for obtaining contact range, and imaging: SAR and ISAR. Radar transmits short microwave pulses to measure contact range from elapsed time between transmission to reception of the echo. The latest systems also obtain target images. Further radar history discussion is found elsewhere [1.33–1.34].

1.4.9 Radio waves > 100cm

Long wave radio can travel worldwide, providing long-distance telegraphy. Medium Frequency (MF) waves provide night-time sky wave transmissions, around 500 miles, relying on total internal reflection from charged ionospheric layers for propagation. High Frequency (HF), around 200 miles, is used for daytime sky wave broadcasts, operating over a wider bandwidth (range of frequencies) than MF. All sky wave transmissions are strongly subject to ionospheric changes.

1.5 Remote sensing application sensing bands

Many bands are now used as a consequence of the new generation of high spectral and spatial resolution satellites. Common nomenclature is used here.

1.5.1 Coastal aerosol band (0.43–0.4μm)

For coastal, bathymetric and aerosol studies. This band detects blue and violet penetrating up to 20–30m in clear water from satellites like Landsat 8 and WorldView-3, for applications such as counting whale populations; bathymetry in coastal water and ocean colour observation; marine vegetation signature detection, such as seagrasses; aerosols that are sensitive to clouds, smoke and haze, and filtering clouds in imagery processing.

1.5.2 Blue (0.45–0.51μm)

Visible blue can reach 20–30m for deep water imaging, penetrating coastal areas and underwater reefs, for water turbidity and sediment, submerged aquatic vegetation and bathymetric mapping. This band is sensitive to atmospheric haze and detects smoke plumes because shorter wavelengths are scattered more by smoke. Clouds are separated from snow and rock, as haze is sensitive to blue and red.

1.5.3 Green (0.53–0.59μm)

Ocean colour sensors look at surface chlorophyll-a concentrations, namely: Sea-viewing Wide Field-of-view Sensor (SeaWiFS), MODIS-Aqua or the MEdium Resolution Imaging Spectrometer (MERIS). The band covers the chlorophyll reflectance peak from leaves and algae. Green discriminates rainforest cutting and broad vegetation classes and distinguishes plants. Algal and cyanobacterial blooms are detectable as clear water reflects weakly across the spectrum with maximum reflectance. When algal blooms are present, green has maximum reflectance.

1.5.4 Red (0.64–0.67μm)

Tropical soils, the built environment and geological features contribute to spectral signature. Red is used in formulas like the Normalised Difference Vegetation Index (NDVI) because plants absorb NIR and red. Bare soil reflectance depends on its composition – for example, Australian soils mimic Martian soils because of their redness. Soils rich in iron oxide, with rust colour, have a high red reflectance. In built environments, we can discriminate man-made objects and vegetation. Spectral signatures for man-made features, such as roads, are best detected in this band. Chlorophyll absorption causes healthy vegetation to reflect more NIR and green than other wavelengths. It absorbs more red and blue and helps quantify vegetation.

1.5.5 Yellow (0.585–0.625μm)

This is used on satellites such as WorldView-2, and provides fine detail between 0.585 and 0.625μm. Tree crowns can suffer from diseases, which gave rise to a Yellow Index (YI) to measure leaf chlorosis. This band can delineate invasive grass and classify tree species and crop types.

1.5.6 Red edge (0.705–0.745μm)

This lies between the NIR and red. For chlorophyll, the red band strongly absorbs light while the NIR generates strong reflection. Transition between these two bands is the red edge. The vegetation response from the red edge is often greatest for chlorophyll content and leaf structure in terms of Leaf Area Index (LAI). Applying this to precision farming discriminates healthy crops from diseased ones and distinguishes crop types.

1.5.7 Near InfraRed 1 NIR1 (0.76–0.9μm)

NIR reflectance is the best way to classify healthy vegetation. Separating classes like water and vegetation is easy, because in the NIR healthy plants reflect well, while NIR is absorbed by water. Vegetation health and biomass content are determined by an internal structure of healthy chlorophyll, which reflects NIR. When plants die, NIR reflectance drops rapidly. Archaeological sites are found by interpreting small differences between crop marks, vegetation, soil and geology. NDVI is used in spectral signatures to measure plant health, and extracts vegetation differences better than if we looked at visible green only.

1.5.8 Near InfraRed 2 NIR2 (0.86–1.0μm)

This delivers more sophisticated vegetation analysis than NIR1 as it is less affected by the atmosphere. Since water is a strong NIR absorber, vegetation reflects strongly. It is easy to delineate forest fragments and quantify forest losses/gains, and to see land/water boundaries.

1.5.9 Cirrus (1.36–1.38μm)

This is specific to cirrus cloud detection. The atmosphere absorbs almost all this band. High-altitude clouds, invisible in other bands, are detected and picked out. Cirrus clouds reflect brightly with land appearing dark.

1.5.10 Short Wave InfraRed 1 SWIR1 (1.5–1.78μm)

SWIR1 discriminates dry from wet soils for geology and soil classification, penetrating thin clouds, smoke and haze better than visible. It discriminates between snow, cloud, ice, water clouds and aerosols, see NIR Meteosat Second Generation (MSG) Channel 3. SWIR1 is sensitive to soil and vegetation moisture content. Reflectance decreases as water content increases, helping distinguish wet from dry earth and mineral deposit detection. SWIR1 sees through smoke to underlying terrain, helping map forest fires effectively. Exploration studies show that spectral signatures indicate minerals like carbonates, ammonium and sulphates or hematite in rocks more easily, mapping rocks and minerals such as with the ASTER SWIR band.

1.5.11 Short Wave InfraRed 2 SWIR2 (2.08–2.35μm)

This band is similar to SWIR1 but is primarily used for imaging soils, geological features and minerals like copper. It is sensitive to vegetation and soil moisture.

Snow and ice feature and clouds appear dark. Water has stronger absorption using SWIR and helps signature responses for monitoring blue-green algae blooms and turbid water. SWIR reflectance is affected by leaf water content, ideal for crop water stress and irrigation practices. SWIR provides greater spectral signature for clay minerals and discriminates between kaolinite types.

1.5.12 Water vapour detection channels (5.35–7.15µm and 6.85–7.85µm)

MSG Channels 5 and 6, and the MSG IR CO_2 band 12.24–14.45µm are sensitive to water vapour levels, as is the panchromatic band.

1.5.13 Panchromatic (0.50–0.68µm)

This band collects all visible light in one channel. It collects a lot of light at once, so spatial resolution provides sharper contrast than collecting red, blue and green separately. Landsat 8's panchromatic band has 15m resolution. Other bands have 30m resolution, except the thermal band, which is 120m.

1.5.14 Thermal InfraRed Sensor TIRS1 (10.60–12.51µm)

This band sees heat. Landsat's thermal IR band detects emitted rather than reflected radiation. It has a coarse 120m resolution and monitors surface temperature and volcanoes. Despite its coarse resolution, TIRS1 can detect and estimate volcano lava discharge rates. Night-time images give energy flux estimations for volcano hazard identification. TIRS also monitors surface temperatures within cities. Urban heat from parks, open water and natural vegetation is coolest, while industrial areas are warmest. A metropolitan area is warmer than surrounding rural areas due to human activities. NOAA's Geostationary Operational Environmental Satellite (GOES) collects thermal IR to understand clouds and for weather prediction.

1.6 Some common remote sensing terms and units

Several fundamental remote sensing terms and units are relevant to different radiation sources.

1.6.1 Radiation sources

Electromagnetic sources, whether natural or artificial, emit wavelengths such as shown in figure 1.4:

(a) A broad continuum of radiation wavelengths, such as the sun

(b) Radiation within a narrow (single spectral) band, such as LED output

(c) Radiation of a single wavelength, such as maritime LIght Detection And Ranging (lidar).

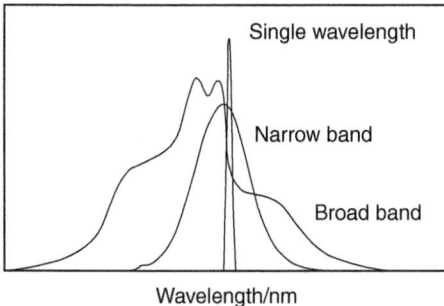

Figure 1.4: *Source output spectra.*

1.6.2 Types of remote sensing imagery

Alongside sources is the question of spectral detail. Recording reflected and emitted radiation is important and is subdivided into several categories. Imagery involves panchromatic, 'pan-sharpened' panchromatic, narrow-waveband, multispectral and hyperspectral.

Panchromatic systems These cover the entire visible. Such broad-band systems usually occupy a large portion of the NIR, providing monochrome black-and-white reflectance in a single band or channel.

Pan-sharpened systems These combine high-resolution panchromatic detail with coloured lower resolution optical bands. Pan-sharpening is a technique that merges high-resolution panchromatic data with medium-resolution multispectral data to create a multispectral image with higher-resolution features. Commercial satellites like GeoEye and IKONOS provide three or more relatively 'coarse resolution' multispectral bands with a finer spatial resolution panchromatic band. There is a compromise in combining the higher spatial resolution of pan images with the medium resolution of multispectral bands, and it is used to compensate for the spectral/spatial compromise in satellite imaging.

Narrow-waveband sensing Target radiation is recorded in a single band. Most objects reflect or emit energy over broad wavelength ranges. There is normally a peak wavelength at which the maximum amount of energy is reflected or emitted. Objects are best differentiated from their backgrounds by measurements made at their radiation peaks. **Bispectral sensing** simultaneously records radiation in two non-adjacent narrow bands, which are compared. The Nimbus THIR and SCMR performed such operations, with 8 and 10 micron bands to estimate SST.

Multispectral imagers These capture data within several specific wavelength ranges across the electromagnetic spectrum. Wavelengths are separated by filters or instruments sensitive to particular wavelengths – for example, Landsat 4 and 5 have three visible, three NIR and one thermal band, combined in different ways. Remote sensing is concerned with compilation of spectra derived from multispectral data, and interpretation by comparison with unique signature 'fingerprint' banks associated with particular minerals, such as gold.

Hyperspectral imagers These capture data within specific wavelength ranges, such as NASA's AVIRIS or Hyperion. Hyperion has 220 spectral bands at 20nm intervals (0.4–2.5 microns) with 30m spatial resolution.

1.7 Resolution issues

Resolution is discussed further in Chapter 2 but key issues are introduced here, namely:

1. Spatial (what area we look at and to what extent, ie how detailed)

2. Spectral (what bands or wavelengths)

3. Temporal (time of day/year/seasons), and revisit periods

4. Radiometric (techniques for measuring radiation, with optics to characterise the radiation power distribution, as opposed to photometric techniques, which characterise light interaction with the eye).

1.8 Attenuation and radiation propagation

Atmospheric attenuation and propagation are discussed further in Chapter 4.

1.8.1 Attenuation

This results from particle scattering suspended in the atmosphere and is related to wavelength, concentration, particle diameter, atmospheric optical density and absorption loss.

1.8.2 Scattering

In practice, complications arise where Earth's atmosphere lies between a sensor and a target. Although wave speed is hardly affected by the atmosphere, the medium greatly affects propagation including: (i) direction, (ii) intensity, (iii) wavelengths reaching the bottom of the atmosphere and (iv) radiant energy spectral distribution (figure 1.5).

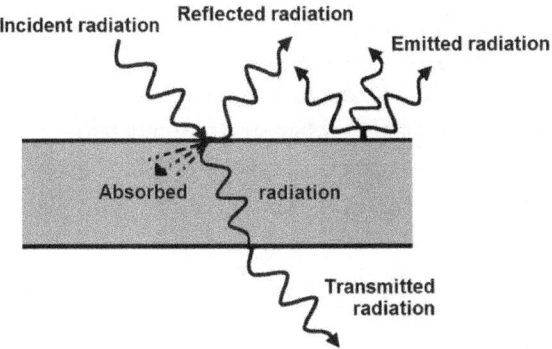

Figure 1.5: *Radiation interactions.*

Frequently, radiation direction and intensity are changed by certain atmospheric particles, which redirect radiation en route through the scattering medium. An understanding of scatter is needed for appropriate sensor selection, and where atmospheric image degradation may be reduced by best wavelength choice.

1.8.3 Absorption

In the atmosphere, radiation absorption takes place in transit. Three gases – water vapour, carbon dioxide and ozone – are efficient absorbers. **In**coming **sol**ar radi**ation** (insolation) is the most important natural source for passive remote sensing, which is attenuated through the atmosphere. The combined effects of absorption and scattering are expressed as an extinction coefficient, which evaluates energy reaching Earth's surface compared with that incident at the atmosphere's outer limits, yielding transmittance. Transmittance decreases as the combined effects of absorption and scattering accumulate.

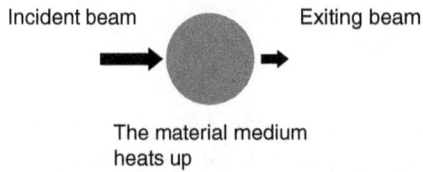

Figure 1.6: *Medium absorption.*

1.9 Albedo

Albedo, from the Latin word meaning 'whiteness', is the fraction of incident sunlight a surface reflects, a useful reflectivity measure for different terrestrial surfaces. It is the insolation percentage reflected back to space. A white body is a perfect reflector – it absorbs no radiation. Hence a white surface temperature stays unchanged. A black body is a perfect radiation absorber, absorbing all energy incident upon it. Typical albedos as a percentage of incident short-wave radiation are given in table 1.3 [1.33].

Fresh snow	0.9
Dense, clean and dry snow	0.86–0.95
Mature marine ice	0.4
Fine sandy soil	0.37
Green grass	0.25
Bare soil	0.17
Pine forest	0.14
Open ocean	0.06
Fresh asphalt	0.04
Great Salt Lake, Utah	0.03

Table 1.3: *Typical albedo values.*

1.10 Spectral radiant flux, reflectance, absorbance and transmittance

1.10.1 Spectral radiant flux $\Phi(\lambda)$

Is the total incident power striking a surface, dependent on wavelength.

1.10.2 Reflectance ρ

Is the percentage of power reflected from a surface, given by multiplying reflectance ρ as a function of wavelength by spectral radiant flux = $\rho(\lambda) \times \Phi(\lambda)$.

1.10.3 Absorbance a

In the atmosphere, transmitted energy is attenuated or absorbed. The radiation percentage absorbed, due to the absorbance $a(\lambda)$, depends on composition, thickness and wavelength. A target may behave differently at different wavelengths – for example, it may absorb in the visible yet be transparent in the infrared (such as germanium).

Absorbance (or a variant *absorptance*, more commonly used with respect to terrestrial surfaces) helps choose the best sensing wavelength. Total power absorbed at a particular wavelength is found by multiplying absorbance (or absorptance) **a(l)** by spectral radiant flux = $a(\lambda) \times \Phi(\lambda)$. Most absorbed energy is converted to heat so temperature increases, making heated surfaces into secondary radiation sources. As solar radiation peak intensity is in the visible, and the Earth/atmosphere system is not a black body, its temperature is less than the sun. Earth's peak emission is in the infrared, emitted from its surfaces day and night, affected little by atmospheric particles such as smoke. If we remotely sense through suitable atmospheric 'windows' and under cloud-free conditions, more is learned than is possible from conventional photography.

1.10.4 Transmittance τ

Of the atmosphere is defined as the ratio of radiation at distance *x* to incident radiation, as a function of wavelength, $\tau(\lambda)$. Total power transmitted at any wavelength is given by multiplying transmittance τ by spectral radiant flux = $\tau(\lambda) \times \Phi(\lambda)$.

Total power is the sum of reflectance, absorbance and transmittance and equals the incident spectral radiation (radiant flux):

$$\Phi(\lambda) = \rho(\lambda) \times \Phi(\lambda) + a(\lambda) \times \Phi(\lambda) + \tau(\lambda) \times \Phi(\lambda) \qquad \text{(eq 1.1)}$$

or simply:

$$\rho(\lambda) + a(\lambda) + \tau(\lambda) = 1 \qquad \text{(eq 1.2)}$$

However, a surface transparent at one wavelength may be opaque at others. The relationship between reflectance, absorbance and transmittance varies across the spectrum. Visible wavelengths are short, so most surfaces reflect light diffusely regardless of incident angle. Some returns reflect in the source direction. Longer wavelengths, such as microwaves, create specular reflection off surfaces, reflecting generally *away* from the source (figure 1.7).

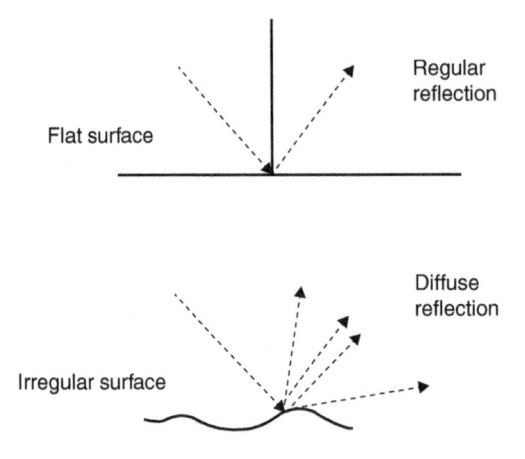

Figure 1.7: *Regular and diffuse reflection.*

At **low** incident angles, reflected energy may be more than combined absorbed and transmitted energies, that is, $\rho(\lambda)>>\alpha(\lambda) + \tau(\lambda)$, while at **high** incident angles, more energy is transmitted, giving detailed surface and subsurface characteristics.

1.11 Radiation emission

The temperature scale has an SI unit called the Kelvin (K not °K). The Celsius scale is often used, as 1 degree Celsius change is equivalent to 1 Kelvin change. To convert from Celsius to Kelvin, add 273.15 – for example, 10°C = 273.15 + 10 = 283.15K.

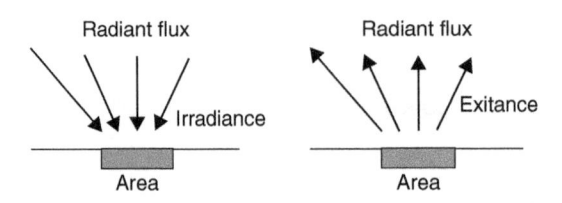

Figure 1.8: *Irradiance (E) and Exitance (M).*

Several important radiant quantities are listed in table 1.4.

Name	Symbol	Equation	Units
Radiant energy	Q	N/A	Joule (J)
Radiant energy density	W	$W = dQ/dv$	$(J \cdot m^{-3})$
Radiant flux	Φ	$\Phi = dQ/dt$	Watt $(J \cdot s^{-1})$
Radiant flux density at a surface	E (Irradiance) into the surface M (Exitance) out of the surface	$E = d\Phi/dA$ $M = d\Phi/dA$	Watt·m^{-2}
Radiant intensity	I	$I = d\Phi/d\Omega$	Watt·steradian^{-1}
Radiance	L	$L = dI/(dA \cos\theta)$	Watt·steradian^{-1}·m^{-2}

Table 1.4: *Several important radiant quantities.*

1.11.1 Irradiance

(E) – incoming radiant flux per unit area (watts per square metre) is the rate photons strike a surface, the amount of energy delivered per unit time (figure 1.8).

1.11.2 Exitance (emittance)

(M) – outgoing radiant flux per unit area (watts per square metre). Note: these may **not** be equal. For example, if irradiance is 1Wm^{-2} and 30 per cent of the irradiance is not reflected **or** emitted, exitance will be 0.7Wm^{-2}. Radiance includes a cosθ dependence for non-normal (off-vertical) illumination (figure 1.8).

All bodies with temperatures above **Absolute Zero** (0K or −273.15°C) emit or exchange energy with their surroundings. Each source emits a characteristic radiation, and for real radiators this output is complex. A **black body** is a theoretical body that absorbs the entire energy incident on it, yet reflects none, emitting with perfect efficiency. The black body concept helped establish laws from which real radiator behaviour is assessed. The **Stefan-Boltzmann law** defines the relationship between total emitted radiation in Wm^{-2} and Temperature T measured in Kelvin (K).

$W = \sigma T^4$ where σ is the Stefan-Boltzmann constant = 5.6697×10^{-8}Wm^{-2}K^{-4}.

Total black body emissive power is proportional to the fourth power of absolute temperature (figure 1.9). Real objects, like the sun, deviate from the ideal, and passage through the atmosphere modifies measured radiance. Emissivity is discussed further in Chapter 3.

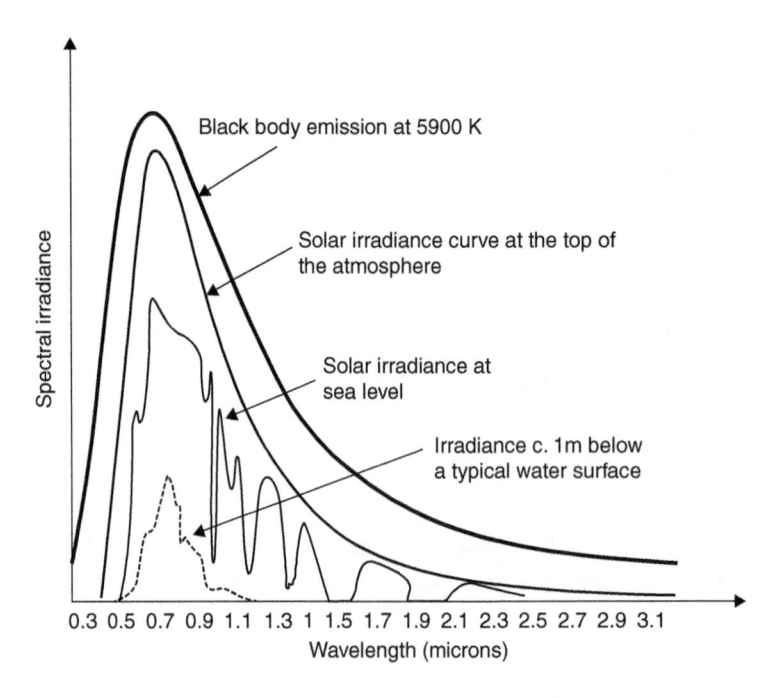

Figure 1.9: *Black body spectra, solar irradiance at the top of the atmosphere and at sea level, and 1m below a typical water surface.*

Hot radiators emit much more energy per unit area than cooler ones; as the temperature increases, total energy emitted also increases.

1.11.3 Kirchhoff's law

As no real body is a perfect emitter, exitance is always less than a black body's. The ratio of emitted/absorbed radiation is the same for all black bodies at the same temperature.

ε = W/W$_b$ where **ε** is the emissivity, **W** surface emittance and **W$_b$** the black body emittance at the same temperature.

Black body emissivity = 1, Perfect reflector = 0. Between these limits, real radiator 'greyness' is given – for example, human skin has emissivity close to a perfect black body = 0.98. Real radiator emission curves vs corresponding black bodies show black body curves are smoother and simpler than observed emissions.

1.11.4 Wien's (displacement) law

Peak black body radiant exitance (λ_{max}) is inversely proportional to absolute temperature (T) thus: $\lambda_{max} = C_3/T$ where $C_3 = 2897\mu mK$. As temperature increases, emitted radiation peak wavelength shifts towards shorter wavelengths. For example, compare the sun 6000K $\lambda_{max} = 0.5\mu m$, with Earth 300K $\lambda_{max} = 9.0\mu m$.

Total solar energy radiated varies widely across spectral bandwidths, table 1.5.

Spectral region	Wavelength range in microns	Percentage of total energy
Gamma and X-rays	10^{-11} m <	Insignificant
Far ultraviolet	0.01–0.2	0.02
Middle ultraviolet	0.2–0.3	1.95
Near ultraviolet	0.3–0.4	5.32
Visible	0.4–0.7	43.5
Near infrared (most reflected)	0.7–1.5	36.8
Middle + Far infrared (most emitted)	1.5–1000	12.41
Microwave and Radio	>1000	Insignificant

Table 1.5: *Spectral region and percentage of total solar energy distribution.*

1.12 Hemispherical reflectance

To describe the angular nature of reflectance, the terms hemispherical and directional reflectance are used. Hemispherical refers to an angle of collection of radiant flux over a hemisphere, and directional refers to collection in one direction only (figure 1.10).

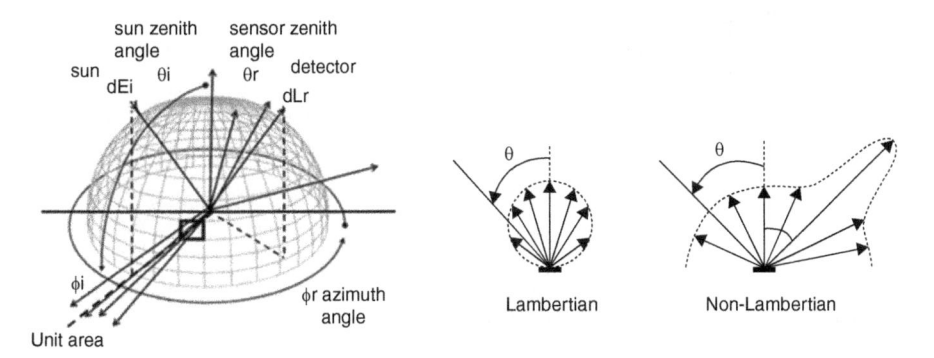

Figure 1.10 : (left) *Hemispherical reflectance through a solid angle;* (right) *Lambertian and non-Lambertian surfaces.*

Most terrestrial spectral reflectance measurements are bihemispherical (angles of incidence and collection are hemispherical) and performed in the laboratory, or bidirectional (where the angles of incidence and collection are directional), as is the case with satellite measurements. In reality, bidirectional reflectance (*brf*) is a ratio between spectral radiance L(λ) at angle θ to the surface and the reflectance from a diffuse reflector at angle θ is used and approximates satellite bidirectional reflectance measurements. Vegetation reflects light unevenly in different directions and is called a 'non-Lambertian' surface, which depends on leaf shape, canopy and vegetation. Viewing angle and surface 'canopy' have a big impact on the brf – for example, a snowy forest with 40–70° incident viewing angle has low red brf, decreasing from 0.15 to 0.05 respectively, while grassy flat farmland increases from 0.15 to 0.17 respectively.

1.13 A remote sensing system

A remote sensing system uses light or suitable bands possessing four key components: a source, its interactions with Earth's surface and with the atmosphere, and a sensor.

1.13.1 Source: the source may be natural, such as the sun, or man-made, such as radar (figure 1.11).

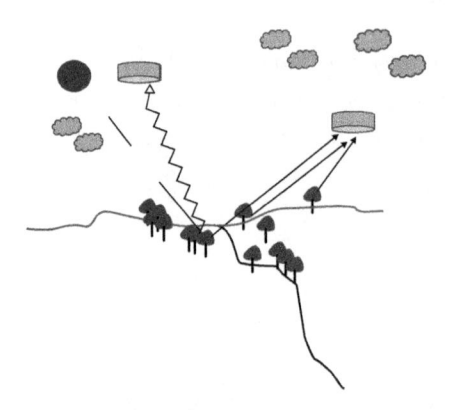

Figure 1.11: *Remote sensing scene.*

1.13.2 Earth's surface interactions

The amount and radiation spectrum emitted or reflected from Earth's surface depends upon surface characteristics.

1.13.3 Atmospheric interactions

Energy passing through the atmosphere is scattered or absorbed. Some solar energy is scattered but reaches Earth's surface, adding to direct irradiance. Some diffuse sky irradiance never reaches the surface but is backscattered towards space.

1.13.4 Sensor

Radiation that has interacted with Earth's surface and the atmosphere is recorded by a sensor, such as a radiometer.

Before remotely sensed images are interpreted, knowledge of *how* radiation interacts with Earth's surface is required. The next few chapters discuss the radiation interactions with the main components of a scene, namely: vegetation, soil, rock and minerals, water, snow and ice, and urban. Urban man-made areas contain complex spatial detail composed of vegetation, soil and water.

1.14 The remote sensing process

The remote sensing process starts from the decision of what must be imaged, with justification, including sensor choice, and passes through several phases: data acquisition to application, figure 1.12. Sensor choice may consider passive high-resolution optical for a small area, but is expensive. SAR can provide all day/night and through cloud sensing, but is active. A cheaper medium resolution passive wide area optical system *might* be preferred but will still be limited to daylight.

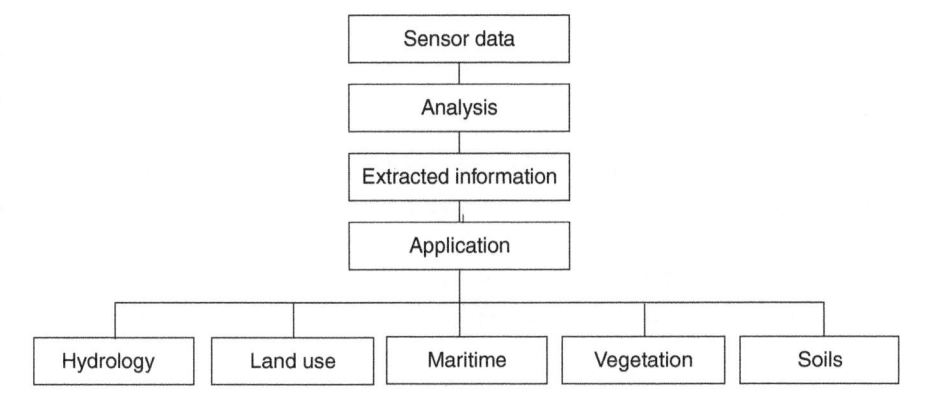

Figure 1.12: *Typical remote sensing work flow diagram.*

Starting from sensor data acquisition, the analysis process is applied to raw data, and includes a mixture of human and machine-based learning, with orthorectification, which removes perspective (tilt) and relief (terrain) effects to create a planimetric corrected image. The resulting orthorectified image has a constant scale so features represent 'true' positions. Extraction of information helps us solve problems in various applications. The typical approach involves interrogation of Earth features, acquisition of sensor data, extracting required information via analysis and interpretation for applications such as hydrology, land or maritime, and is considered a looped cycle, figure 1.13.

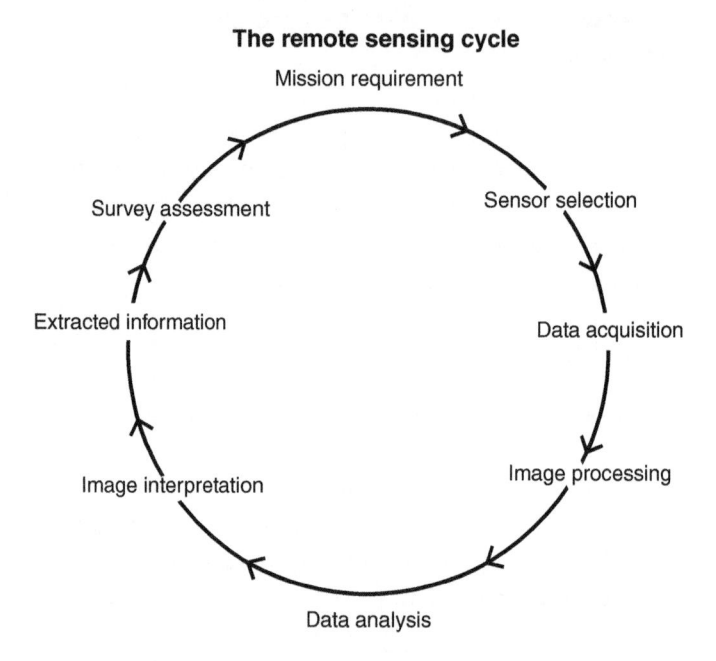

The remote sensing cycle

Mission requirement → Sensor selection → Data acquisition → Image processing → Data analysis → Image interpretation → Extracted information → Survey assessment → (Mission requirement)

Figure 1.13: *The remote sensing cycle.*

Questions

1.1 Describe two key historical developments that took place in remote sensing, and explain why you consider them to be important.

1.2 Discuss what effects refraction, absorption and scattering have on waves travelling between two atmospheric layers, and what may happen to waves on reflection at Earth's surface. Consider the size of particles in the atmosphere.

1.3 With careful use of internet sources, discuss in some detail two systems for Active maritime remote sensing and two systems for Passive maritime remote sensing. Consider parameters such as: range, power, imaging/non-imaging etc.

1.4 State the relationship between incident spectral radiant flux, reflectance, absorbance and transmittance. If total incident spectral radiant flux is 100W, ρ = 0.4 and α = 0.2, what is the reflected power? (2 significant figures). What is transmittance (1 decimal point) and transmitted power (2 significant figures)?

1.5 Explain the reasons for remote sensing's widespread use today.

1.6 If a sandy beach's peak emission wavelength = 9.73 microns, what is the surface temperature in Kelvin (K) and Celsius (C)?

1.7 If the intensity of a thermal source is proportional to T^4, by how much will radiated intensity increase if temperature increases by 10 per cent?

1.8 Find the UV dose absorbed by a skin surface in Jm^{-2} if intensity at the sea level skin surface is 280Wm^{-2}, over 1 hour. Note: UV dose = intensity × time in seconds. Comment on the likely relationship between UV dose and changes to DNA.

1.9 A visible blue wavelength travels between two sheets of glass. If glass reflectance = 0.9, what is reflected light intensity recorded after **five** reflections if $I_0 = 1$ Wm^{-2}?

1.10 The Skylab S–192 scanner had 13 channels. Comment on suitability of these four bands for possible *chlorophyll-a algal detection*:

Channel 1: 0.41–0.46 microns   Channel 3: 0.52–0.56 microns
Channel 8: 0.98–1.08 microns   Channel 13: 10.2–12.5 microns

REFERENCES

[1.1] www.overshootday.org/

[1.2] data.unicef.org/topic/child-survival/under-five-mortality/

[1.3] www.avoid.uk.net/2016/03/the-impact-of-weather-extremes-on-agricultural-commodity-prices-f3/

[1.4] www.zerohedge.com/news/2017-05-01/wheat-soars-most-record-after-freak-snowstorm-blanket-midwest

[1.5] www.uasvision.com/2017/02/09/uk-coast-guard-studying-uav-options/

[1.6] 'Zimbabwe: Shattered lives – the case of Porta Farm', Amnesty International and Zimbabwe Lawyers for Human Rights, Summary: 31 March 2006, AI Index: AFR 46/004/2006.

[1.7] 'Application of satellite imagery to monitoring human rights abuse of vulnerable communities, with minimal risk to relief staff', CR Lavers, Baroness Cox et al., XV *Journal of Physics: Conference Series*, 178 012039 (Bristol IOP Publishing, 2009).

[1.8] 'Rapid NDVI assessment in land clearance studies using high-resolution satellite imagery', CR Lavers and T Mason, Proceedings of the Remote Sensing and Photogrammetry Society Annual Conference (2011).

[1.9] 'High-resolution IKONOS satellite imagery for normalized difference vegetative index-related assessment applied to land clearance studies', CR Lavers and T Mason, *Journal of Applied Remote Sensing*, 11(3), 035008 (4 August 2017).

[1.10] *Introduction to Environmental Remote Sensing*, 2nd Edition, EC Barrett and LF Curtis (Routledge, London, 1999, ISBN 0748740066).

[1.11] *Remote Sensing: Principles and Interpretation*, 3rd Edition, Floyd F Sabins (W.H. Freeman and Company, San Francisco, 1996, ISBN 9780716724421).

[1.12] 'Comparison of satellite-derived with ground-based measurements of the fluctuations of the margins of Vatnajökull, Iceland, 1973–92', Richard S Williams Jr et al., *Annals of Glaciology* Vol. 24, 1996, pp.72–80.

[1.13] pubs.usgs.gov/gip/topomapping/topo.html

[1.14] [1.11]

[1.15] [1.11]

[1.16] unitar.org/unosat/

[1.17] Military uses of space, www.parliament.uk/documents/post/postpn273.pdf

[1.18] Defense Standardization Program Case Study, December 2006, Number 273,

share.ansi.org/Shared%20Documents/Other%20Services/SBB/Case-Study-NAVSTAR-GPS.pdf

[1.19] en.wikipedia.org/wiki/Radarsat-1

[1.20] 'Nanosatellites for Future Naval and Maritime Applications', Christopher Lavers, 2017 blogs,plymouth.ac.uk/dcss/2017/07/13/nanosatellites-for-future-naval-and-maritime-applications/

[1.21] 'Tomorrow's Nano-satellites', Christopher Lavers and Paolo Quaranta, Military Technology Journal online (21 April 2017).

[1.22] Applications of Earth Observation – Surrey Satellite Technology, www.sstl.co.uk/Downloads/Brochures/SSTL-Applications-Brochure-Web.

[1.23] www.nasa.gov/cygnss

[1.24] 'A Dynamical Theory of the Electromagnetic Field', J Clerk Maxwell, *Philosophical Transactions of the Royal Society*, London, 155, pp.459–512, published 1 January 1865.

[1.25] *Physics Wave Concepts for Marine Engineering Applications*, CR Lavers and S-K Lavers (Bloomsbury Publishing, London, 2017, ISBN 1472922151).

[1.26] 'An Overview of Infrared Remote Sensing of Volcanic Activity', Matthew Blackett, *Journal of Imaging* 3(2), (12 April 2017), p.13.

[1.27] 'IR SAMS down but not out', P Donaldson, *Defence Helicopter* (December 2003/January 2004), pp.23–26.

[1.28] 'The Technology and Applications of Thermal Imaging', CR Lavers, *Electronics and Beyond*, No. 126 (June 1998), pp.30–35.

[1.29] 'FLIR for Safety', B Dagiliatis, *Defence Helicopter*, September 1996, pp.40–45.

[1.30] *Essential Sensing and Telecommunications for Marine Engineering Applications*, Christopher Lavers (Bloomsbury Publishing, London, 2017, ISBN 1472922182).

[1.31] 'Non-destructive high-resolution thermal imaging techniques to evaluate wildlife and delicate biological samples', CR Lavers et al., *Journal of Physics: Conference Series 178, No.1, Sensors and Their Applications XV* (October 2009) (Proceedings doi: 10.1088/1742–6596/178/1/012040 ISSN: 1742–6588).

[1.32] 'What heat can reveal – using thermal imagery for wildlife research', CR Lavers, *The Wildlife Professional* (Winter 2009), pp.66–68.

[1.33] Reeds Marine Engineering and Technology Series, Vol 15: *Electronics, Navigational Aids and Radio Theory for Electrotechnical Officers*, S Richards (Bloomsbury Publishing, London, 2013, ISBN 9781408176092).

[1.34] *Stealth Warship Technology*, Reeds Marine Engineering and Technology, Vol 14, Christopher Lavers (Thomas Reed, London, 2012, ISBN 9781408175255).

2
Visible, Near Infrared and Ultraviolet Electromagnetic Radiation Interactions at the Earth's Surface

'But everything exposed by the light becomes visible, for it is light that makes everything visible.' Ephesians 5 v 13 New Testament

2.1 The interaction of visible, NIR and ultraviolet with Earth's surface

In the next few chapters we will treat the interactions of visible, near infrared and ultraviolet wavelengths, thermal wavelengths and microwave wavelengths separately. In practice, complex spectral interactions exist – for example, short wavelength light can be absorbed and re-radiated by surfaces at longer thermal or microwave wavelengths. We will examine specific surfaces: water, vegetation, soil, rock and minerals, snow and ice, which represent the main natural challenges presented to maritime and land Earth observation today. The concept of spectral radiant flux $\Phi(\lambda)$ was introduced in Chapter 1. Flux incident on Earth surfaces are reflected $\rho(\lambda)\Phi(\lambda)$, absorbed $\alpha(\lambda)\Phi(\lambda)$ or transmitted $\tau(\lambda)\Phi(\lambda)$. In reality there is a combination, but through conservation of energy none is lost in these processes:

$$\Phi(\lambda) = \rho(\lambda)\Phi(\lambda) + \alpha(\lambda)\Phi(\lambda) + \tau(\lambda)\Phi(\lambda) \quad \textbf{(eq 2.1)}$$

Or simply:

$$1 = \rho(\lambda) + \alpha(\lambda) + \tau(\lambda) \quad \textbf{(eq 2.2)}$$

Consequently, the spectral radiant flux proportions of reflected, absorbed or transmitted $\tau(\lambda)$ are *dissimilar* for different surfaces – for example, we observe a white surface as having high reflectance at all wavelengths, while green plants have high reflectance in visible green only. The proportion of spectral radiant flux can change for a surface as a function of wavelength and across widely separated bands: visible, thermal or microwave. Reflectance variations, however, depend upon four interrelated factors:

(1) Sensor radiometric resolution, or **Sensitivity**, (2) Atmospheric **Scattering**, (3) Surface **Roughness** and (4) Spatial Reflectance and Resolution **Variations** in recorded scenes.

Sensor **sensitivities** vary significantly in ability to detect radiance differences. For example, the Landsat Thematic Mapper (TM) radiometer on Landsat 4–5 satellites detects 256 radiance levels (8 bit quantisation or 2^8 Digital Number DN values); the MultiSpectral Scanner (MSS) radiometer, however, detects only 64 levels, with 2^6 DN levels, while the MOderate Resolution Imaging Spectroradiometer (MODIS) has 2^{12} DN values. Differences easily distinguished by MODIS are less easily distinguished by TM, and may be undetected by MSS.

Atmospheric **scattering** has an unfortunate result of decreasing overall radiance otherwise received by sensors, resulting in reduced contrast between a surface and background. This makes it harder to detect surfaces clearly in mist or fog. Terrestrial viewing doesn't generally require atmospheric corrections needed for satellite observation.

Surface **roughness** is vital as surfaces must be rough enough to allow radiation to interact. If a surface is smooth, radiation is reflected without scattering, so little energy is transmitted to sensors or echo-based detectors. However, most of Earth's surface appears rough at visible and NIR wavelengths. To determine *how* rough a surface is requires quantification as a function of wavelength, using *Rayleigh's* surface roughness *criterion*.

Surface variation *below* $\lambda/(8\cos\theta)$, where λ is the wavelength and θ the incident angle, is considered **smooth** [2.1]. Surfaces appearing equally rough (or smooth) from all angles are **Lambertian**. Most terrestrial features, however, are not equal and are thus non-Lambertian. Viewing angle must be specified – for example, with 10cm wavelength and 60° incident angle, $\lambda/(8\cos\theta) = 2.5$cm. With mean surface roughness = 2cm, a surface is considered smooth.

Spatial **variations or variability** is the fourth factor as sensors record radiance not just from a targeted ground area but also from nearby surrounding regions.

Spatial reflectivity is an important factor and affects recorded signal strength. It was estimated by Forster [2.2] that only 52 per cent of recorded radiance for each pixel in satellite imagery originates from the intended or 'viewed' ground pixel. Although for large tracts of 'featureless' ocean or monocultural farming this presents little problem in misreading radiance (except perhaps edges of such regions), it causes large problems in urban environments where rapidly changing land cover spatial variability – roads, parks, houses, pavements etc – results in fluctuations of surrounding pixels, table 2.1. Recorded response is affected by cover surface in centre and surrounding pixels – for example, reflection from a 100 per cent grass centre pixel surrounded by concrete neighbours appears due to 50 per cent grass: 50 per cent concrete.

Estimated percentage reflectance	Band 4	Band 5	Band 6	Band 7
House	16.0	18.0	25.0	30.5
Road	15.9	14.2	11.1	10.7
Concrete	22.6	28.3	34.3	38.6
Tree	5.9	1.5	19.3	33.3
Grass	14.6	12.0	29.1	43.8
Other	16.9	12.0	14.7	15.0

Table 2.1: *Estimated percentage reflectance for various types of urban surface cover: Other – building roofs, high-density residential, commercial and industrial (after Forster [2.2]).*

For better understanding of surface interactions, we will take components of visible light and NIR radiation, then shorter ultraviolet wavelengths, looking at respective interactions with water (hydrosphere), vegetation and chlorophyll, soil, rocks and minerals, and snow and ice (cryosphere). This is, of course, a crude approximation of the full range of surface interactions.

2.2 Water properties

Water properties are considered briefly here.

Sea water physical and chemical characteristics: oceans and seas cover two-thirds of Earth's surface. A 1 atmosphere pressure increase (1 bar = 10^5 Nm^{-2}) occurs for 10m descent in sea water, so marked pressure changes occur in typical ascent or descent. Buoyancy depends on water Specific Gravity (SG), which for sea water at the surface is 1.022 but at 9km depth is 1.070, a small buoyancy change. Horizontal movement is caused by tides and wind, and vertical movement by convection currents. Tides are generated by gravitational attraction to bodies in space, the principal bodies being the moon and sun. Water mixing stabilises temperature and salinity, and results in plankton movement. The highest temperature is 30.5°C (the Red Sea), the lowest −20°C (the Southern Ocean), but most waters are below 5°C. Aquatic photosynthesis is mostly by diatoms in near surface layers (above 168m where there is sufficient light). Sea water has a constant pH of 8, is slightly alkaline, acting as a buffer against pH change, especially those introduced by pollutants. Salt water contains common ions in concentration order: sodium > magnesium > calcium > potassium. Sea water is saturated with oxygen (34%), nitrogen (65%) and CO_2 (0.3%), with water concentration similar to air concentration.

Fresh water physical and chemical characteristics: Lakes vary in depth (the deepest is Lake Baikal, Russia, 1,642m), and there is much water movement variation. Ponds are largely unaffected by wind and may be stagnant with vegetation, while streams flow to the sea. Fresh water buoyancy has lower SG than sea water. Pressure has little effect on shallow water. Movement is important in lakes and streams. Small ponds in winter may have a bottom temperature of 4°C, depending on water clarity (turbidity); in clean water, one can see beyond 180m. Fresh water has 100 times *less* total salt content than sea water. pH in fresh water can undergo large changes (it is a poor buffer solution) and varies with dissolved CO_2. pH for aquatic life occurs between 4.6 and 8.5.

Air and satellite photographs monitor quality, quantity and distribution of water resources. We are primarily concerned with water pollution detection, lake eutrophication assessment and flood estimation. The basic properties of sunlight interaction with clear water are considered in this chapter. Most light entering 'clear' water is absorbed within 2m of the surface, dependent on wavelength. Reflected IR is absorbed in tens of centimetres, resulting in dark tones in even shallow water on images as there is almost no IR reflection from water, giving sharp water/land boundaries, see figure 2.1 and plate section.

Figure 2.1: *Isles of Scilly NIR imagery (2017), Courtesy Dr T Mason, Channel Coast Observatory. CASI SW UK Isles of Scilly NIR image, with red strong NIR vegetation reflections, dark (low returns) even from shallow water.*

Visible spectrum absorption varies dramatically with the particular water body. For bottom detail photography in clear water, the best light penetration is between 0.48 and 0.6μm. Blue penetrates well but is strongly scattered and underwater haze results. Red penetrates a few metres.

2.3 The interaction of visible light and NIR with water (the hydrosphere)

The liquid water and ice absorption spectra from the UV into the mid IR is similar but diverges above 10 microns. In brief, water absorbs long wavelength visible and NIR more than ice. Clear water appears blue-green due to strong reflectance at short wavelengths, and dark when viewed at red or NIR. Suspended upper water layer sediments increase reflectivity and water can appear brighter, shifted to longer wavelengths. Suspended sediments may be confused with shallow clear water. Chlorophyll in algae absorbs more blue and reflects green, making water appear greener with algae present. Water surface topography (rough or smooth) can confuse interpretation. Radiant incident water flux is largely absorbed or transmitted, absorbing NIR strongly (figure 2.2).

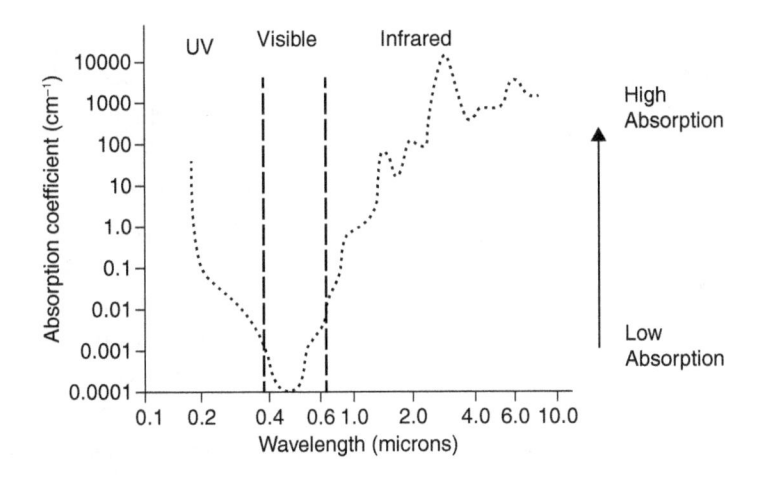

Figure 2.2: *UV, visible, IR water absorption. Absorption of electromagnetic radiation by sea water (after Wolfe and Zissis [2.3]).*

After undergoing absorption, little NIR radiation is left to reflect or transmit ($\alpha \gg \rho + \tau$), giving sharp contrast between water and land boundaries, such as a CASI airborne NIR image (figure 2.1).

Three factors affect water spatial reflectance: **depth**, suspended **materials** and surface **roughness**. In shallow water, some radiation is reflected from the bottom and top water surface. In shallow pools or streams, *underlying* material impacts overall reflectance – for example, a painted blue swimming pool makes water 'appear' blue. Common materials suspended in water include non-organic sediments, like tannin and chlorophyll. Non-organic silts and clays increase visible scattering and reflectance, especially in *littoral* environments, those near sea or lake shores [2.4]. Tannin is the main colour arising from decomposing humus (yellow to brown), giving **decreased** blue and **increased** red reflectance. Water with chlorophyll has reflectance resembling vegetation with increased green, decreased blue and red reflectance. However, chlorophyll content must be high before changes are detected even with multispectral sensitive satellites, such as SeaWiFS. Surface roughness also affects reflectance; smooth surfaces reflect light specularly, with low reflectance back to sensors, but if rough there is increased surface scatter, increasing reflectance. Various sensitive ocean colour sensors exist: MODIS, SeaWiFS or MERIS, with multiband sensor performance [2.5].

2.4 Underwater light attenuation

The deeper you go in marine waters, the darker it becomes. Light levels drop or attenuate with depth rapidly; even in clear waters, blue-green levels reduce to 1 per cent of surface values in 100m. Red light intensity falls faster, dropping to 1 per cent in just 10m. Intensity falls with depth across the visible and NIR through **attenuation** by two different mechanisms: **absorption** and **scatter**. Conditions for vision underwater on a good day compare with fog above water. It isn't specifically low intensity that prevents distant viewing, but scatter from fog water droplets, which reduce contrast below the minimum to identify objects against the background. Absorption varies in amount and spectral distribution across the oceans, depending on water quality. Water molecules have broad absorption resonances in the visible and IR; however, selective absorption also occurs. Blue-green propagates best in sea water, selectively absorbing low energy (long wavelength) red, so low absorption of high-energy short wavelength blue generates water's characteristic colour. The visible absorption coefficient $A(\lambda)$ for pure water is given in table 2.2, as are $B(\lambda)$ the scattering coefficient and $C(\lambda)$ the total loss coefficient.

λ nm	Colour	A m^{-1}	B m^{-1}	C = (A + B) m^{-1}
410	Violet	0.016	0.007	0.023
470	Blue	0.016	0.004	0.020
535	Green	0.053	0.002	0.055
555	Yellow-green	0.06	0.002	0.069
575	Yellow	0.094	0.002	0.096
600	Orange	0.244	0.001	0.245

Table 2.2: *Absorption, scattering and total loss coefficients.*

Absorption is at its minimum at the short wavelength blue end. The complex distribution of sea water dissolved salts have little effect on overall optical properties, but have different and significant *corrosive* properties, a critical factor when considering dissolved salt compositions in the various world's oceans upon different immersed metal alloys. Using instruments more sensitive than the human eye, satellites measure ocean colour and track sea surface temperatures and currents. Mapping ocean colour reveals the presence and concentration of marine animals, phytoplankton, sediments and dissolved organic chemicals. The Sea-viewing Wide Field-of-view Sensor (SeaWiFS) provided qualitative data on global ocean bio-optical properties (figure 2.3, see plate section). Such images tell us much about an ocean's state. Mariners use remote sensing satellites to measure various parameters and provide ocean surveillance on a global scale. Each data point represents an individual unit of Earth's surface. New satellites and sensors mean there is an increasing amount of data collected.

The visible sea water scattering coefficient $B(\lambda)$ m^{-1} is given (table 2.2); as expected, for λ^{-4} Rayleigh-type scattering, $B(\lambda)$ is minimum in the red and maximum in the blue. It is the high blue scatter combined with low blue absorption that together yield $C(\lambda)$, so:

$$C(\lambda) = A(\lambda) + B(\lambda) \qquad \textbf{(eq 2.3)}$$

The direct consequence of absorption and scattering intensity falls to a level *below* that predicted from considering A or B alone and also due to the inverse square law.

Example 2.1: For water at 470nm, find C if $A = 0.016$m^{-1}, and $B = 0.004$m^{-1}:

$C = 0.016 + 0.004 = 0.020$ m^{-1}

Suppose a narrow collimated light beam, of radiance L, propagates a short distance dr of water (figure 2.4). Over this distance there is intensity loss dL, so incident intensity (Lin) attenuates to Lout = (L-dL).

We write beam Transmittance T = radiance out/radiance in

So T = Lout/Lin = (Lin − dL)/Lin, and

T = 1 −dL/Lin

Attenuance A, or loss per unit length, is given by: A = dL/Lin where Lin = L so T = 1-A

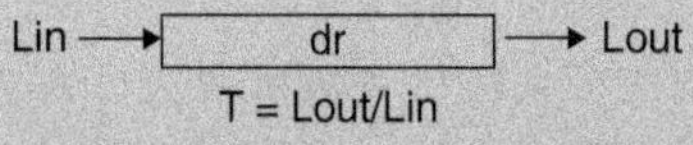

> Lin ⟶ [dr] ⟶ Lout
> T = Lout/Lin
>
> Figure 2.4: *Transmittance in terms of radiance changes.*
>
> Attenuance doesn't specify the water distance over which radiance attenuates. So a beam attenuation coefficient, c, is defined as the attenuance of an infinitely thin water layer divided by layer thickness.
>
> $C = -A/dr = -dL/(L \times dr) = -1/L \times dL/dr = -d/dr\,(\ln(L))$ **(eq 2.4)**,
>
> where C is the rate of change of natural logarithm of radiance intensity with distance in m^{-1}.

2.4.1 The Beer-Lambert law

A law developed for above water optics describes underwater light logarithmic attenuation. For radiance it is derived from (**eq 2.4**) as follows:

$C = -d/dr \times \ln(L)$ **(eq 2.5)**

Considered over distance dr and integrating both sides between 0 and R:

$$\int_0^R C \times dr = -\int_0^R (d/dr \times (\ln[L])) \times dr$$

Therefore $-C \times \int_0^R dr = \int_0^R d/\ln[L]$

So:

$-C \times R = \ln\dfrac{[L(R)]}{[L(0)]}$

$-C[r]_0^R = [\ln L]_0^R$

Thus: $-C \times R = \ln[L(R)] - \ln[L(0)]$

Taking antilogs: $e^{-CR} = L(R)/L(0) = Lout/Lin$

So $L(R)/L(0) \times e^{-CR} = T$, and $L(R) = L(0) \times e^{-CR}$ **(eq 2.6)**

2.4.2 The optical distance (γ)

The exponent CR in the Beer-Lambert law is geometrical distance × beam attenuation coefficient, and termed the optical length γ = C × R. A given optical length of water results in a specific attenuance.

> **Example 2.2**: Light attenuates equally when propagated through 5m with c = 0.2m^{-1} or through 20m with c = 0.05m^{-1}. What is γ? In each case γ = 5 × 0.2 = 20 × 0.05 = 1.0 and transmittance T is given by:
>
> T = L(R)/L(0) = exp^{-CR} = exp$^{-γ}$ = exp$^{(-1)}$ = 0.37 = 37%.

2.4.3 The beam transmissometer

Beam transmittance T is measured with a beam transmissometer. The beam attenuation coefficient, C, is calculated from T.

L(R) / L(0) = T = e^{-CR} So −CR = ln[T], or

C = −ln[T]/R   (**eq 2.7**)

Ocean transmissometers contain a light source, an optical system that produces a narrow optical beam and a detector. A photon can strike a detector without aqueous interaction or is absorbed in water, or scattered by particles and lost before reaching the detector. Beam attenuation is partly due to absorption and scattering, so:

C(λ) = A(λ) + B(λ) = −ln[T(λ)]/R(λ)   (**eq 2.8**)

The Beer-Lambert law for radiance may be expressed as:

L(r) = L(0) × exp[-C × r] so

  = L(0) × exp[-(A + B) × r]   (**eq 2.9**)

The beam transmissometer was developed 30 plus years ago at Oregon State University and measures transmission with a pulsed light source and detector. A receiver is open to ambient background light, so a transmitted beam is pulsed and ambient light level measured. The Alphatrack MkII, a common transmissometer from Chelsea Instruments, is available at four wavelengths: 470, 565, 590 and 660nm and three path lengths: 5, 10, or 25cm.

2.4.4 Diffuse attenuation coefficients and the Beer-Lambert irradiance law

A broad field of diffuse irradiance, E, such as a narrow beam, is attenuated diffusely by the Beer-Lambert law. Irradiance decreases with depth z, so:

$E(z) = E(0) \times \exp[-K \times z]$ (**eq 2.10**)

K is the diffuse attenuation or extinction coefficient. Photons follow longer paths due to scatter, which increases absorption *probability*. Turbidity increase causes greater scattering and B increases. Photons scattered out of a directional beam are lost, while photons scattered within a field of diffuse light aren't necessarily lost, but may be absorbed. Small turbidity increases results in large visibility decrease but only a small decrease in overall light intensity, as observed with fog or light smoke.

2.4.5 Optical refraction underwater

Light refracts at interfaces between two different optical media. Refractive index, N_{sw} of sea water, increases with increasing density, salinity S, wavelength λ, pressure p, and decreasing temperature T. Table 2.3 shows refractive index for different wavelengths in distilled water at 10°C:

Wavelength (nm)	Refractive index (n)
226.5	1.39422
361.05	1.3487
404.41	1.34389
589	1.33408
632.8	1.33282
1013.98	1.32591

Table 2.3: *Refractive index of distilled water at 10°C [2.6].*

Austin and Halikas (1976) tabulated N(S,T,p,λ) [2.7]. N changes slightly over ocean ranges of S, T, p and λ. Small N_{sw} changes are due to random thermal fluctuations and salt ion molecular agitation.

2.5 The interaction of ultraviolet with water (the hydrosphere)

The sun is the greatest UV radiation source in the terrestrial environment. UV radiation at Earth's surface varies widely, depending on local atmospheric and environmental conditions. Under clear skies, the most important parameters are solar zenith angle (time of day and latitude), total ozone content, cloud cover, amount and aerosol type, altitude above sea level and ground [2.8]. The influence of altitude shouldn't be underestimated. With every 1km increase in altitude, UV increases 10–12 per cent. The UV spectrum divides into three regions: UVA, UVB and UVC, absorbed by atmospheric ozone, water vapour, oxygen and carbon dioxide. UVA is **not** filtered much by Earth's atmosphere, but UV is weakly reflected from Earth's hydrosphere. As seen later, UV-stimulated fluorescence is a major detection tool for oil on water. Water generally reflects below 10 per cent of incident UV. However, at 0.5m depth UV levels are 40 per cent of that at the surface. UVA can be a threat to the health of ship's crew and passengers exposed on long sea voyages even on cloudy days, as UV is attenuated little. Cloud can enhance UV levels because of enhanced scatter. Outdoor activities and sunbathing may result in excess UV exposure. Raised awareness and lifestyle changes are needed to protect against this.

Long wavelength UVA (315–400nm) accounts for around 95 per cent of UV radiation reaching Earth's surface. It penetrates deep skin layers and is responsible for immediate tanning, contributing to skin ageing and wrinkling. Recent studies suggest it may promote skin cancer development. Medium wavelength UVB (280–315nm) is biologically active but cannot penetrate beyond skin layers. It is responsible for delayed tanning and burning, short-term effects contributing to skin ageing and promoting skin cancer development. Most solar UVB is filtered by the atmosphere.

Short-wavelength UVC (100–280nm) is the most dangerous UV radiation, but it is completely filtered by the atmosphere and doesn't reach Earth's surface. Extreme UV below 121nm ionises air *so* strongly it is absorbed well before it reaches the ground.

Ozone is an effective UV radiation absorber. As ozone thins, the atmosphere's protective filter progressively reduces. Ozone and the atmosphere are discussed in Chapter 5. If ozone were condensed to a liquid and spread evenly over Earth, the layer would be around 4 microns thick. The ozone amount is expressed in

Dobson units (DU), with a 'normal' amount about 300 DU. People are exposed to higher UV levels where there is ozone depletion, especially UVB. Ozone depletion is caused by man-made chemicals released into the atmosphere and will continue until the use of chlorine and bromine compounds is drastically reduced. However, the lifespan of chemicals already released will cause ozone depletion to persist for years to come. A full recovery of ozone is not expected until 2050. Depletion of emissions of chlorofluorocarbons (CFCs), methyl bromide (CH_3Br), nitrogen oxides (NOx) and other substances released by human activities are required. Continuous observations since the 1980s show that ozone has decreased by 3–6 per cent, resulting in a 6–14 per cent increase of UVB radiation at Earth's surface. CFC half-lives range from 50 to 150 years, so each CFC molecule may destroy many ozone molecules. It is estimated that a 1 per cent decrease in ozone results in a 1.1 per cent weighted UV irradiance increase [2.9]. Liquid water has its absorption minimum near the UV-visible boundary [2.10].

2.6 Vegetation

About 70 per cent of Earth's land surface is covered by vegetation, and photosynthetic activity takes place in near surface regions of the world's oceans (the *photopic zone*). Here we look at visible, NIR and UV interactions with vegetation and chlorophyll.

2.6.1 The interaction of visible and NIR with vegetation

Vegetation reflectance varies with wavelength. To understand *why* vegetation reflects some wavelengths more than others, consider a typical leaf made of structured organic fibres: pigments (dyes), water-filled cells and air spaces. Each of these three features – pigment, physical structure and water content – affect reflectance, absorbance and transmittance [2.11], see figure 2.5.

Although leaves have high NIR, photosynthesis rates are driven by exposure to red (680nm) and FIR (700nm). Emerson [2.12] showed that exposure to both 680 and 700nm produces enhanced photosynthesis rates. Leaf canopy wetness and dryness affects overall reflectance.

2.6.2 Visible pigment absorption

Most marine environment or land plants contain four main characteristic pigments, which absorb visible light during photosynthesis: chlorophyll A and B, β-carotene and xanthophyll. Chlorophyll A and B are the key pigments, absorbing blue and red.

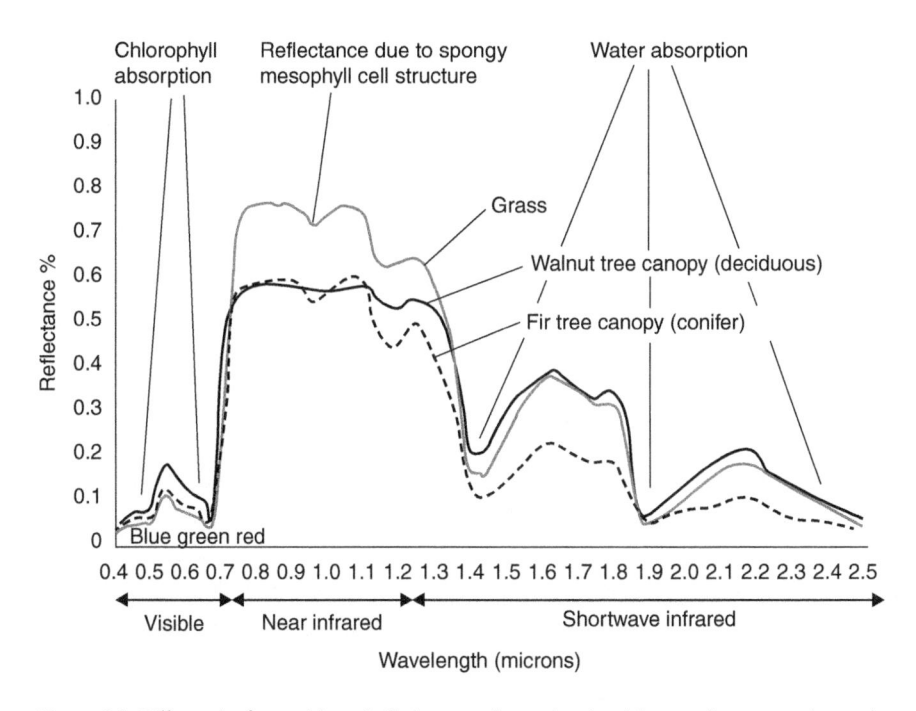

Figure 2.5: *Different leaf type chlorophyll pigment absorption: low blue, medium green, low red and high NIR reflectance.*

Chlorophyll A absorbs at 0.43 and 0.67µm, while chlorophyll B absorbs at 0.45 and 0.65µm (figure 2.6a, see plate section). Chlorophyll has absorption in the blue (short wavelength) and goes from a high excited state, via heat loss, to a lower energy state. There is also red absorption, the lowest excited state, followed by heat loss and fluorescence.

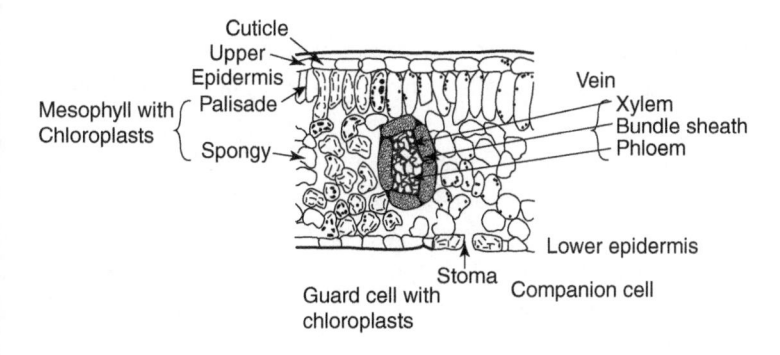

Figure 2.6b: *Typical leaf structure.*

β-carotene and xanthophylls absorb blue to green [2.13]. Chlorophyll A plays the key role in photosynthesis, but plants have pigments that participate in photosynthesis called *antenna* pigments, which don't excite chlorophyll A; however, absorbed energy is passed to the chlorophyll A and used in photosynthesis. When plants photosynthesise faster than they use up photosynthetic products in respiration, they exceed the compensation point – that is, light levels exceed where photosynthesis is greater than respiration. At high intensities plants grow, add to reserves, and reproduce. However, below this point in winter without shedding leaves, deciduous trees use up reserves and eventually die.

2.6.3 Leaf structure and NIR reflectance

Leaf structural boundaries determine overall NIR reflectance, between membranes and cytoplasm in the upper leaf surface and cells and air spaces in the spongy lower half [2.14]. A combination of leaf pigments and physiological structure, due to destructive interference between surface reflections, also plays a part. NIR reflections give healthy green leaves characteristic **low red** and **blue**, **medium green** and **high NIR** reflectance. Reflectance differences between terrestrial species depends on thickness, affecting pigment content and structure, figure 2.6b. For example, mature oak leaves absorb much radiation and transmit little, freshly budded oak leaves absorb less and transmit more, while marine kelp 'forests' have moderate absorption with 'weak' physiological structure, as buoyancy removes the need for xylem.

Forest vegetation canopies are composed of individual leaves, different plant structures, background and shadow. Measured sensor reflectance depends on leaf area within a canopy, the **Leaf Area Index** (**LAI**), the leaf area per unit ground area. Grasslands have canopies with high LAI values. Reflectance depends on soil background, vegetation decay, solar and sensor elevation, canopy geometry and canopy changes. On land, soil has a big effect on observed reflectance. For example, looking down on rows in a field, more soil is visible than looking sideways across the rows. Dark soil also has a different reflectance compared with vegetation and light soils! It is important to choose the correct wavelength to determine water or vegetation. Leaf reflectance is sensitive to moisture content [2.15], see figure 2.7. Several water vegetation absorption bands exist near 1.4, 1.9 and 2.7µm, and further weak bands near 0.96 and 1.1µm, reducing phytoplankton and leaf reflectance. At some wavelengths, water and vegetation reflectance are similar. Leaf reflectance in water absorption bands *decreases* as water content and thickness *increase*. There are also atmospheric 'windows' largely free of water absorption.

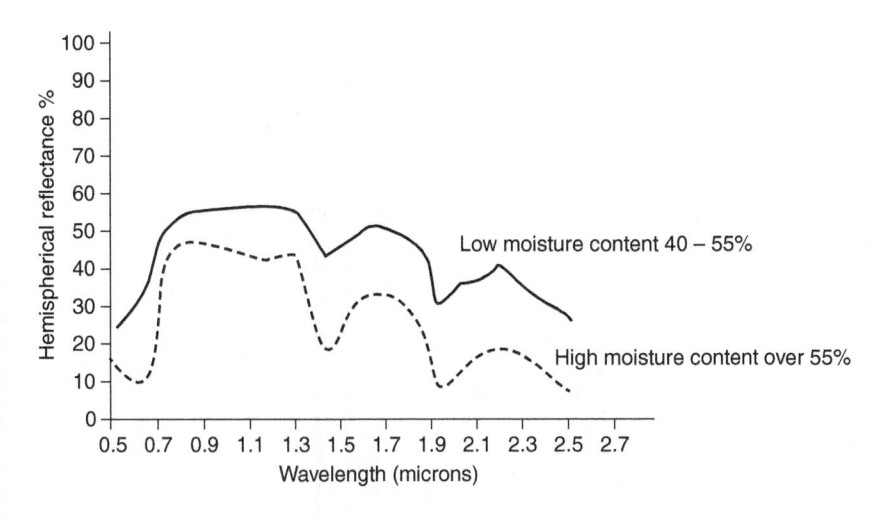

Figure 2.7: *The effect of moisture content in corn leaf reflectance (after Hoffer and Johannsen [2.16]).*

IR reflectance can monitor plant stress, with reflectance falling to low percentage values for certain species. SeaWiFS provided quantitative data on global ocean bio-optical properties in the visible and NIR region.

2.7 Time dependent characteristics

2.7.1 The effect of solar and sensor elevation (height)

Vegetation doesn't reflect equally in all directions, and is non-Lambertian. When the sun is **low**, sky radiation penetrates canopy less (being more a surface effect) and is reflected more from canopy surfaces, so reflectance is **high**, with some reflected down and lost (figure 2.8a, left). When the sun is **high**, overhead radiation penetrates deep below a canopy, and reflectance may be **low**. Waves that travel 'more straight down' to the forest floor pass through a forest volume, may reflect many times, and suffer absorption from forest and ground surfaces, leaving little returned up into space, so reflectance can be surprisingly low – that is, weak returns are observed (figure 2.8). Multiple reflections between trees on the way down (and back) for slightly off direct ground reflections also suffer attenuation (loss).

Sensor elevation angle, whether on satellites or UAVs, determines ground and soil visibility. Elevation angle determines the amount of visible soil. As elevation angle moves off vertical, the result is **less** visible soil (beneath the canopy) and **more** vegetation (reflection from the top surface of the canopy). This is similar to how radar aperture size changes. This angular cosθ effect must be taken into account as satellite sensors move off a direct downward view. Many measurements are vertical but there is increased collection off vertical; for example, Landsat TM views up to 9° while Spot HRV views up to 27°.

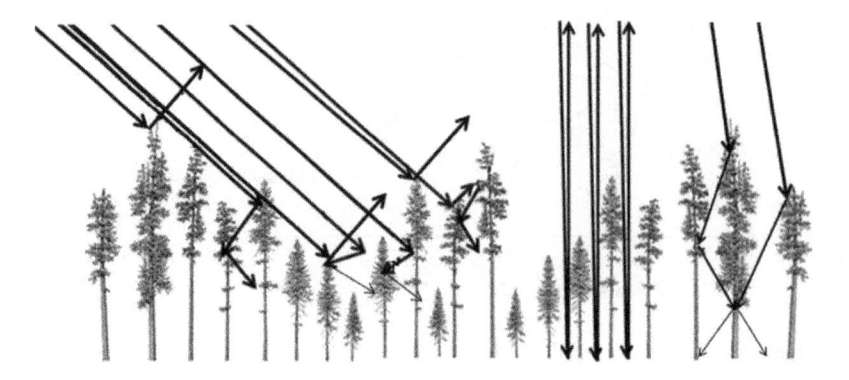

Figure 2.8: *High reflectance surface interactions, low reflectance multiple reflections with vegetation.*

2.7.2 The effect of solar and sensor azimuth

This is largely a question of whether we view towards the sun or away. Given that for most passive systems the source is the sun, it isn't a surprise that canopy reflectance is higher when a sensor looks towards rather than away from the sun. When viewing angle is vertical, solar azimuth effects on reflectance increase with a **decrease** in solar angle and canopy roughness **increase**. At high solar angle, there is almost no shadow. At low incident angle, shadow is greater depending on canopy roughness. If this paragraph were a maize field planted in horizontal rows, with letters representing heads of maize, if the sun was low in the sky and parallel to the rows (left to right) a detector measures a higher soil proportion and *less shadow* reflectance than if the sun were at 90° to the row (top to bottom), see figure 2.9.

Figure 2.9: *Left view along rows, right view across rows.*

2.8 Canopy geometry changes

2.8.1 The effect of vegetation decay or senescence (phenological cycles)

As plants ripen, die and decay through their life cycle, NIR reflectance doesn't decrease much. However, plant pigments break down, increasing blue and red reflectance. Pigment changes due to annual leaf decay are seen in the autumn, which under 'stressed' leaf conditions create colourful yellow to red leaf colours. With non-deciduous canopy (such as grasslands), red reflectance is maximum in autumn and minimum in spring. NIR reflectance tends to be maximum in summer and minimum in winter. These are referred to as *phenological changes*.

2.9 The interaction of UV with vegetation

Plants exploit blue and UVA to drive DNA repair processes. Surprisingly, UV damage can be repaired by subsequent exposure to blue light or UVA. Blue light or UVA exposure activates an enzyme (photolyase) that repairs damaged DNA sequences. Photoreaction is the main defence against plant UV-induced damage. Production of UV-protective *phenolic* (aromatic ring) compounds found in algae and various vascular plants provide defence against injury, infection and stress (frost, high temperatures and drought). UV-absorbing compounds accumulate in the plant surface layers. Presence of these substances in plants is one reason soil is dark brown, as phenolics in dead plant materials eventually form soil. UV absorbing compounds accumulate in leaf epidermal cells and act as selective sunscreens, reducing further penetration of UV into leaf tissue. They don't affect visible light penetration, which is essential for photosynthesis. Insects benefit from seeing flower patterns visible in UV only, while carnivorous plants use their UV markings to attract pollinators to their doom in insect traps! Food grown under glass or picked early, however, does **not** possess UV protective properties.

2.10 Vegetation, and Normalised Difference Vegetation Index

A Vegetation Index (VI) is a spectral combination of two or more bands to enhance vegetation assessment and its properties, and allow reliable spatial and temporal comparisons of terrestrial and marine photosynthetic activity and of canopy structure. There are many VIs; some use the inverse relationship between red and NIR reflectance associated with healthy green vegetation. Vegetation measurement attributes include: LAI, percentage green cover, chlorophyll content, green biomass and absorbed photosynthetically active radiation. VIs specify the spectral bands and the calculation method. With recent hyperspectral remote sensing technology advances, high-resolution reflectance spectrums are available, beside traditional multispectral VIs. VIs have been developed specifically for hyperspectral data, as well as narrow-band VIs.

Green plants absorb solar radiation through photosynthesis. Leaf cells reflect solar radiation in the NIR (just under half of incident solar radiation). Strong absorption at these wavelengths would result in plants overheating and damaging tissues. So although green plants appear dark in the photosynthetic region, they appear relatively bright in the NIR. Early Earth observation instruments, such as NASA's ERTS and NOAA's AVHRR, acquired data in the visible and NIR to exploit strong plant reflectance differences and determine spatial distributions on satellite images. NDVI is calculated using red and NIR reflectance ratios and varies between −1.0 to +1.0, and is similar to NIR/red ratios. NDVI is directly related to photosynthetic capacity and plant canopy energy absorption.

In our own high-resolution satellite imagery work [2.17] we use:

NDVI = $(\rho_{NIR} - \rho_{Red})/(\rho_{NIR} + \rho_{Red})$ (**eq 2.13**), analogous to Landsat's NDVI. NDVI is usually applied to vegetative land cover, or a normalised temporal imagery method $C_T = (\rho_{After} - \rho_{Before})/(\rho_{After} + \rho_{Before})$ (**eq 2.14**), where ρ is the satellite camera's digital number. Historically, military aerial surveillance operations used single band image comparison. However, a better choice would subtract before NDVI from after NDVI. NDVI has a normalised index with modular values of 0 to 1 but isn't radiometrically calibrated, and is one of several normalised indices for vegetation, soil and other common surfaces. The NOAH AVHRR instrument has detection capabilities between 0.55 and 0.7, and 0.73–1.0 microns. Each AVHRR pixel is derived from 1 km² of land surface using an equation based on the principle that healthy photosynthetic vegetation absorbs strongly in the visible, while reflecting strongly in the NIR.

NDVI = $(\rho_{NIR} - \rho_{Visible})/(\rho_{NIR} + \rho_{Visible})$ (**eq 2.15**)

Example 2.3: A dense vegetative index may have 65% NIR reflectance and only 8% visible reflectance. Find the NDVI %.

Thus NDVI = (65−8)/(65 + 8) = 39%

Vegetation extraction and tree mapping from IKONOS imagery is possible with a high spatial resolution index. Other enhanced vegetation indices are also used with automated forest canopy characterisation.

While snow may show a low ratio, reflected NIR and reflected visible are both similar: NDVI = (0.33 − 0.28)/(0.33 + 0.28) = 0.08. Lake water chlorophyll distribution can be estimated with Landsat TM data [2.18].

2.11 The interaction of visible and NIR with soil

Most solar flux incident on soil is reflected or absorbed and little transmitted. Most soil reflectance is similar, increasing with wavelength, figure 2.10.

From equation (2.2), for a fixed wavelength $1 = \rho + \alpha + \tau$ and soil τ approaches zero then $1 \approx \rho + \alpha$ or rearranging: $\alpha \approx 1 - \rho$ ... **(eq 2.16)**

Example 2.4: If the hemispherical reflectance of a dry sandy loam soil is 0.2 at 0.5 microns, what is the soil's absorbance?

As a function of wavelength $\alpha(\lambda) \approx 1 - \rho(\lambda)$,

then $\alpha \approx 1 - 0.2 = 0.8$

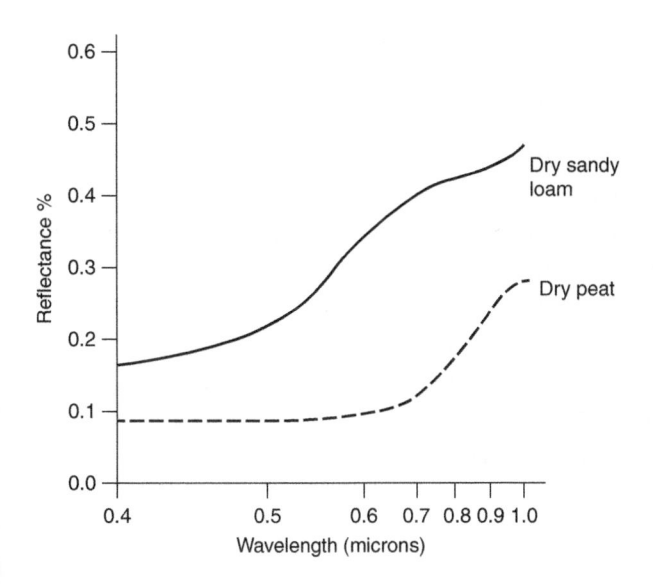

Figure 2.10: *Reflectance from different soil types.*

Soil has five key factors that determine reflectance, given in order of importance: (i) moisture content, (ii) organic content, (iii) iron oxide, (iv) structure and (v) texture, providing an easy acronym: **MOIST**! These factors are interrelated; for example, texture (proportion of sand, silt and clay) relates to structure (physical arrangement of sand, silt and clay), and soil's ability to hold moisture (water). Wetting of all soils reduces measured reflectance, figure 2.10.

2.11.1 Soil moisture, structure and texture

Clay soils have a *strong* water-holding structure, resulting in *rough* ploughed surfaces, with *high* moisture content and *low* diffuse reflectance. In contrast, sandy soils have *weak* structure, leading to *smooth* ploughed (crumbly) surfaces, with *low* moisture content and *high* (often specular) reflectance. In the visible soil, moisture **reduces** reflectance.

2.11.2 Organic matter and iron oxide

Soil organic matter is visibly dark and decreases reflectance up to 4–5 per cent organic content. If organic content is above 5 per cent, soil appears 'black' and no further increase in organic reflectance is perceived. Iron oxide makes soil red. Iron oxide selectively absorbs green 0.5–0.6μm and reflects red 0.6–0.7μm [2.19]. The red/green reflectance ratio aids location of haematite iron ore deposits from satellite data and accounts for the redness of the planet Mars from recent Curiosity rover data [2.20].

2.11.3 The interaction of ultraviolet radiation with soil

Reflected UV from soil and vegetation provides information about the material it is reflected from. There are few studies conducted on soil and vegetation spectra in the UV, visible and NIR combined, except this one [2.21]. Absorbance of soils in wet and dry conditions are shown, figure 2.11a.

This is similar but different to vegetation absorbance in fresh and dried conditions, figure 2.11b. Terrestrial UV sensing at close range and space-based UV sensing from satellite of clouds can retrieve cloud parameters from satellite-based reflectance measurements in the UV and oxygen A band [2.22].

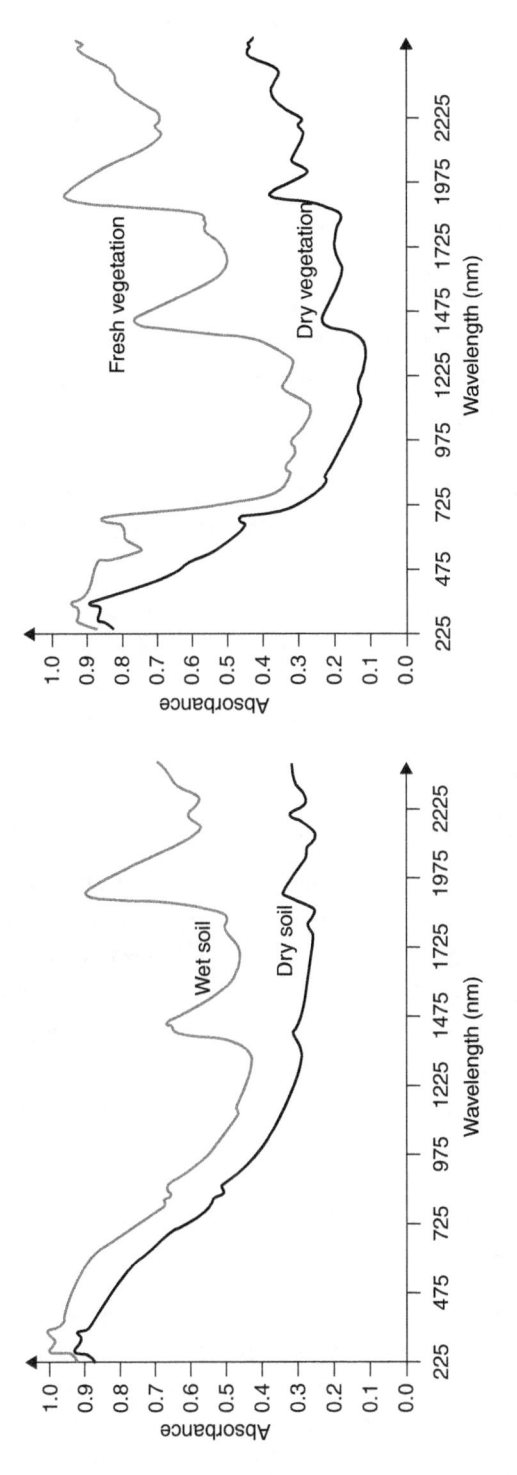

Figure 2.11a (left): Absorbance of wet and dry soil; Figure 2.11b (right): Absorbance of fresh and dry vegetation (after Bogrekci and Lee [2.21]).

2.12 The interaction of visible, NIR and ultraviolet radiation with rock and minerals

The optical properties of native rock or minerals depend on **how** they interact with light. Several important properties apply to minerals to help identify them:

1. *Colour*: Mineral colour is mainly the result of selective light absorption.

2. *Transparency*: Minerals can be transparent, translucent or opaque. The amount of light absorbed depends on atomic bonds in a rock's chemical structure. Completely transparent minerals have no colour, such as diamond (transparent from the UV-FIR). Opaque rocks absorb or reflect all incident light on them.

3. *Lustre*: A mineral surface property, how it reflects light independent of colour – for example, metallic lustre varies depending on the crystal face, from metallic to glassy.

4. *Refraction and reflection*: Refraction changes the light path, while reflection is light 'bouncing off' a surface. Real and imaginary refractive indices over various wavelengths are available for common terrestrial rocks and glasses 0.3–50 microns [2.23].

2.13 The interaction of ultraviolet with rocks and minerals

Fluorescence: Under UV light, rocks and minerals emit light. Fifteen per cent of minerals produce fluorescence visible to humans. One of the first people credited with observing mineral fluorescence was George Stokes in 1852, who noted fluorite produced a blue glow when illuminated with UV. Minerals temporarily absorb UV and a little later release light at a different wavelength. A photon strikes an electron within a mineral, exciting the electron so it temporarily moves to a higher orbital state. It then falls back to its ground state, losing the stored energy, emitting a photon at a longer wavelength.

2.14 The interaction of visible and NIR radiation with snow and ice (the cryosphere)

Ice cover is the main factor governing the Arctic radiative budget. Monitoring ice cover is of great importance. Light transmission and reflection from ice and snow are used. One of the first studies was Grenfell [2.24], determining spectral albedos (400–1,000nm) and extinction coefficients (400–800nm) for melt ponds, snow and various bare ice types. Albedo was largest in the 400–600nm range, figure 2.12 [2.23] for different kinds of ice, for visible and NIR reflectivity during the ablation period on Peyto Glacier, Canada [2.25]. Spectral time over day reflectivity changes are observed, figure 2.12.

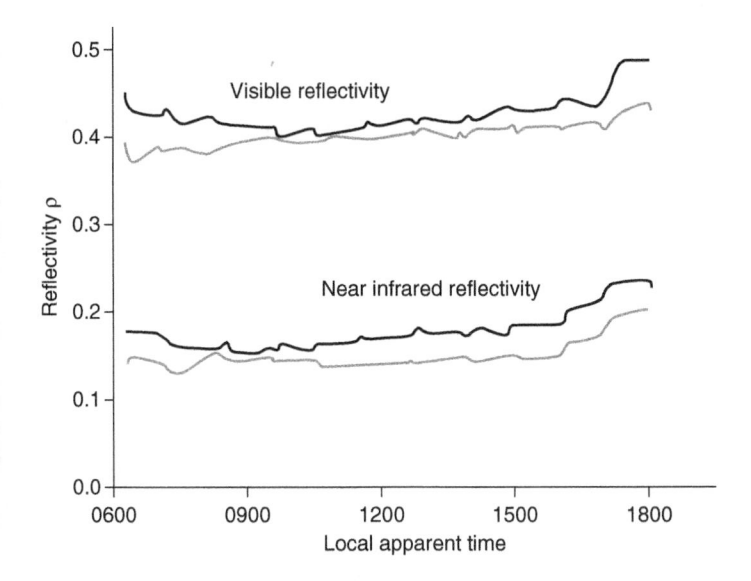

Figure 2.12: *Before and after for visible and NIR reflectance [after Wolfe and Zissis [2.23]). The impact of high-melt conditions showed reflectivity on consecutive days: day 1 (light plot high melt), day 2 (thick plot low melt).*

2.15 The interaction of ultraviolet with snow and ice (the cryosphere)

Snow, to the unaided eye, is scatter-dominated with scattering wavelength independent between 350 and 600nm. Ice exhibits strong UV absorption below 170nm. With increasing wavelength, absorption becomes weak and minimum at 400nm. NIR absorption is moderate, and strong on through the IR 3–150 microns, becoming weak again in the microwave region above 1cm. Liquid water absorption spectrum generally parallels ice from UV to mid-IR, but diverges above 10 microns. Between 300 and 600nm ice absorption is *so* weak ($1 \approx \rho + \tau$) it can be considered zero, and clean fine-grained snow reflects 97–99 per cent of incident light [2.10]. Albedo value is uniformly high (0.96–0.98) across the UV to visible, nearly independent of snow grain size and solar zenith angle. NIR albedo is lower, dropping below 0.15 between 1.5 and 2 microns [2.26].

Depending on the Earth observation feature, we may obtain different responses to absorbance, transmittance and reflectance in the visible, NIR and UV bands. By measuring reflected or emitted surface energy over such wavelengths, we obtain a **spectral response** for Earth's differing land cover. By comparing surface responses we can distinguish between them, something we might otherwise be unable to do if we only looked at one wavelength. Spectral responses vary with time and depend on factors such as leaf greenness, ice reflectivity or geographical location. Knowing *where* to 'look' spectrally and understanding factors that influence spectral response of features is critical to correctly interpreting interaction of radiation with a surface.

Questions

2.1 Incident radiant flux is 100W at 500nm. Reflectance is 0.1. Absorbance is 0.4. What is the transmittance? What are the reflected and absorbed powers?

2.2 What does figure 2.2 show about infrared's suitability for underwater imaging?

2.3 Consider figure 2.10. Draw where you think a mix of dry sandy loam and dry peat would appear. What would *wet* peat reflectance look like? If $\alpha = 0.3$, what would the likely reflectance ρ be?

2.4 Explain with a diagram the components of a remote sensing system and the relevance of spectral radiant flux.

2.5 Compare the reflectance of visible, NIR and UV for surface interactions with water, vegetation, soil and urban surfaces.

2.6 Explain the importance of ocean colour to MODIS and SeaWiFS.

2.7 A non-Lambertian surface is measured with 5cm waves at 20° viewing angle. Average surface roughness = 3cm. Show if the surface is rough or smooth.

2.8 Using table 2.1, discuss discrimination of grass from roads using VI reflectance differences in bands 4–7.

2.9 Express the reflectance of sandy loam and peat at 0.5 microns. Find the ratio of sandy loam/peat in dB. From figure 2.11, what is the difference in reflectance between wet and dry moisture content at 1.4 microns? State this in dB.

2.10 Suppose there is a narrow collimated light beam, with radiance L, propagated through a short distance, dr, of water. Over this distance there is an intensity loss of dL, so incident intensity Lin is attenuated to Lout.

(i) Show how transmittance can be written in terms of L and dL.
(ii) If attenuance A, or loss per unit length, is given by A = dL/L, simplify the transmittance equation.
(iii) The Beer-Lambert law describes underwater radiance logarithmic attenuation ln[L] in terms of beam attenuation coefficient C:
$\quad\quad$ C = −d/dr × ln(L) over a short distance r.
$\quad\quad$ Integrate both sides between 0 and R with respect to r to show the result: L(R) = L(0) × e^{-CR}
(iv) Using your result, calculate the outgoing light intensity if C = 0.2m^{-1} over a 3.1m water path for an incident intensity of 2.7Wm^{-2}.
(v) Find the value of the transmittance T as a percentage.

References

[2.1] *Remote Sensing: Principles and Interpretation*, 3rd Edition, Floyd F Sabins (W.H. Freeman and Company, San Francisco, 1996, ISBN 9780716724421).

[2.2] 'Some Urban Measurements from Landsat Data', Bruce Forster, *Photogrammetric Engineering and Remote Sensing*, Vol. 49, No. 12 (December 1983), pp.1693–1707.

[2.3] *The Infrared Handbook*, WL Wolfe and GJ Zissis (Environmental Research Institute of Michigan for the Office of Naval Research, Department of the Navy, Washington DC, 1978).

[2.4] 'Classification of turbidity levels in the Texas marine coastal zone', EA Weisblatt, JB Zaitzeff, and CA Reeves, Proceedings of the Conference on Machine Processing of Remotely Sensed Data (October 1973), pp.3A.42–3A.59.

[2.5] 'Dynamic range and sensitivity requirements of satellite ocean colour sensors: learning from the past', C Hu, L Feng et al., *Applied Optics* Vol. 51, No. 25 (1 September 2012).

[2.6] 'Water refractive index in dependence on temperature and wavelength: a simple approximation', AN Bashkatov and EA Genina, Proceedings of SPIE 5068, Saratov Fall Meeting 2002: *Optical Technologies in Biophysics and Medicine IV* (13 October 2003).

[2.7] 'The Index of Refraction of Seawater', RW Austin and G Halikas, Visibility Laboratory of the Scripps Institution of Oceanography, La Jolla, CA (January 1976).

[2.8] 'UVR climatology', M Blumthaler and A Webb, in *UV effects in Aquatic Organisms and Ecosystems* (Comprehensive Series in Photochemistry and Photobiology Volume I), EW Helbling and H Zagarese (Eds) (The Royal Society of Photochemistry, Cambridge, 2003), pp.21–58.

[2.9] 'Ozone and ultraviolet-B irradiances – Experimental determination of the radiation amplification factor', M Blumthaler, M Salzgeber, W Amback, *Photochemistry and Photobiology*, 61 (February 1995), pp.159–162.

[2.10] 'Visible and near-ultraviolet absorption spectrum of ice from transmission of solar radiation into snow', SG Warren, RE Brandt, and TC Grenfell, *Applied Optics*, Vol. 45, No. 21 (20 August 2006).

[2.11] 'Remote sensing brightness maps', JR Jensen and ME Hodgson, *Photogrammetric Engineering and Remote Sensing*, 49(I) (1983), pp.93–102.

[2.12] 'Dependence of yield of photosynthesis in long wave red on wavelength and intensity of supplementary light', R Emerson, *Science*, 125 (1957), p.746.

[2.13] *Mechanism of Photosynthesis*, CP Whittingham (Edward Arnold, London, 1974).

[2.14] 'Leaf reflectance of near-infrared', HW Gausman, *Photogrammetric Engineering and Remote Sensing*, Vol. 40(2) (1974) pp.183–191.

[2.15] 'Biological and physical considerations in applying computer-aided analysis techniques to remote sensor data', RM Hoffer in *Remote Sensing the Quantitative Approach*, PH Swain and SM Davis (Eds) (McGraw Hill, New York, 1978) pp.227–289.

[2.16] 'Ecological potentials in spectral signature analysis', RM Hoffer and CJ Johannsen, in *Remote Sensing in Ecology*, PL Johnson (ed.) (University of Georgia Press, Athens, 1969) pp.1–16.

[2.17] 'High-resolution IKONOS satellite imagery for normalized difference vegetative index-related assessment applied to land clearance studies', CR Lavers and T Mason, *Journal of Applied Remote Sensing*, 11(3) 035008 (4 August 2017), doi:10.1117/1.JRS.11.035008.

[2.18] 'Chlorophyll distribution in Lake Kinneret determined from Landsat Thematic Mapper data', M Mayo et al., *International Journal of Remote Sensing*, Vol. 16(1), January 1995, pp.175–182.

[2.19] 'Spectral reflectivity of the main soil types and the possible use of diffuse reflectance in soil studies', AI Obukhov and DS Orlov, *Pochvovedenie*, No. 2 (1964), pp.83–94.

[2.20] 'Magnetic properties experiments on the Mars exploration rover *Spirit* at Gusev Crater', P Bertelsen et al., *Science*, Vol. 305, Issue 5685 (August 2004), pp.827–829.

[2.21] 'Soil particle size effect on absorbance spectra of sandy soils in UV-VIS-NIR regions', I Bogrekci and WS Lee, ASAE Paper No. 043112, St Joseph, Michigan, 2004.

[2.22] 'Retrieval of cloud parameters from satellite-based reflectance measurements in the ultraviolet and the oxygen A-Band', B van Diedenhoven et al., *Journal of Geophysical Research: Atmospheres* Vol. 112(15) (16 August 2007), doi: 10.1029/2006JD008155.

[2.23] 'Optical Properties of Some Rock Samples', JB Pollack et al. (1972) hdl.handle.net/1811/9072.

[2.24] 'The optical properties of ice and snow in the Arctic Basin', TC Grenfell and GA Maykut, *Journal of Glaciology*, Vol 18(80) (1977), pp.445–463.

[2.25] 'Visible and near-infrared reflectivity during the ablation period on Peyto Glacier, Alberta, Canada', PM Cutler and DS Munro, *Journal of Glaciology*, Vol 42, No 141 (1996), pp.333–340.

[2.26] 'Reflection of solar radiation by the Antarctic snow surface at ultraviolet, visible, and near-infrared wavelengths,' TC Grenfell et al., *Journal of Geophysical Research: Atmospheres*, Vol. 99(D9), 20 September 1994, pp.18669–18684.

3
Thermal Sensors

'It was supposed that the rays from the sun and fixed stars could reach the earth through the atmosphere more easily than the rays emanating from the earth could get back to space.'
  John Tyndall, 1859, after having started his experimental work on radiant heat

Introduction

To understand Earth's climatic system, it is vital to monitor sea ice *changes* in polar latitudes. Heat energy exchanged between ocean and atmosphere significantly affects weather systems. Sea ice insulates water from the colder atmosphere above it. When ice melts or upper sea surfaces freeze, they affect upper ocean temperature distribution, which in turn affects current patterns.

3.1 Thermal radiation and its interactions with the Earth's surface

Thermal radiation concepts were introduced in Chapter 1. We looked at Planck's black body equation and the Stefan-Boltzmann and Wien's laws, which state there is a maximum wavelength at which a black body radiates peak output, determined by its temperature.

For the sun, the spectral radiant exitance is given by:

$$M_\lambda = \frac{2\pi h c^2}{\lambda^5} \frac{1}{e^{\frac{hc}{\lambda kT}} - 1} \quad \textbf{(eq 3.1)},$$

where:

M_λ = spectral radiant exitance (emittance) in $Wm^{-2}\mu m^{-1}$,

λ = wavelength in metres, and T the blackbody's temperature in Kelvin (K),

$h = 6.625 \times 10^{-34} Js^{-1}$, $c = 3 \times 10^8 ms^{-1}$, and $k = 1.38 \times 10^{-23} JK^{-1}$.

Example 3.1: What is $M_{\lambda,sun}$ at 10 microns?

Consider emitted solar energy, where we will take the temperature of the solar surface at 6000K for simplicity (5782K exactly: with wavelength λ now substituted in microns:

$$M_{\lambda,sun} = \frac{3.7492 \times 10^9}{\lambda^5} \frac{1}{e^{\frac{14413}{\lambda \times 6000}} - 1} \text{Wm}^{-2}\,\mu\text{m}^{-1} \quad \textbf{(eq 3.2)}$$

$$M_{\lambda,sun} = \frac{3.7492 \times 10^9}{10^5} \frac{1}{e^{\frac{14413}{10 \times 6000}} - 1} \text{Wm}^{-2}\,\mu\text{m}^{-1}$$

$$M_{10\,microns,sun} = 1.38 \times 10^5 \text{Wm}^{-2}\,\mu\text{m}^{-1}$$

but is far from the visible peak wavelength emittance. The solar emittance becomes the incident radiation upon the Earth's outer atmosphere. The Earth of course intercepts only a small component of the total solar output. All Earth surfaces have temperatures above Absolute Zero (0K), emitting and absorbing radiation from their surroundings. Earth has a mean surface temperature around 300K (27°C), re-radiating at peak exitance between 3 and 50 microns.

3.2 Emissivity

Let's first consider how objects emit heat. The emissivity of a surface describes how good an emitter it is. The higher the emissivity, the greater the heat proportion transmitted/emitted through the surface, analogous to visible transmittance. For example: glass Fresnel optical transmittance is 0.96 – that is, 96 per cent of incident light is transmitted. Similarly, if human skin thermal emissivity is 0.98, 98 per cent of body heat reaching the skin surface is emitted through, with 2 per cent reflected back inside the body. Not all of a body's received energy is re-radiated. There are no true 'black bodies', but rather 'grey bodies' that emit a proportion of received energy, expressed by emissivity. High emissivity near 1 indicates an object absorbs and radiates most of the incident energy. A low value (near 0) indicates an object absorbs and radiates a smaller proportion. Table 3.1 gives typical emissivity values.

Material	Emissivity	Material	Emissivity
Open canopy vegetation	0.96	Brick	0.93
Closed canopy vegetation	0.99	Steel	0.16
Pure water	0.993	Granite	0.96
Wet loamy soil	0.95	Dry loamy soil	0.92
Aluminium	0.08	Snow	0.8

Table 3.1: *Typical emissivity values.*

Black body radiation is distributed over a narrow wavelength range (figure 1.11), with power emitted by unit area per unit wavelength interval plotted against wavelength. Overall emission shape is temperature independent, but the maximum emission wavelength increases with decreasing temperature. Simultaneously, the height of the maximum *decreases*, since total energy emitted decreases with falling temperature.

Consider a remote sensing satellite in space. Energy radiated with temperature is proportional to the fourth power of absolute temperature. Satellite temperature is determined by balancing radiation *absorbed* from the sun and that *emitted* from its surfaces, ideally 20–70°C. Neither sun nor satellite is a perfect black body but emitted radiation wavelength dependence for each is similar to that from a black body at the same temperature. All but 3 per cent of incident solar radiation is between 0.3 and 3 microns, while for a surface at 27°C it is 4.8–60 microns. Without this spectral distribution difference, it wouldn't be possible to control satellite temperature by adjusting surface absorption and emission, as at a given wavelength both cannot change independently. Temperature control is possible because $\varepsilon(\lambda)$ and $a(\lambda)$ (or rather the averages of both) vary from one another if measured at different wavelengths. Solar absorbance $a(\lambda)$ is the fraction of incident solar radiation absorbed. Thermal emittance $\varepsilon(\lambda)$ is the fraction of total surface black body radiation emitted. If the absorbing surface body area is Sa and the emitting area is Se (for a rotating cylinder in space, this is approximately equal), we arrive at the following:

$$aS_a E_0 = \varepsilon S_\varepsilon \sigma T^4 \qquad \textbf{(eq 3.3)}$$

where E_0 is the incident energy, giving

$$T^4 = \frac{(S_a E_0)}{(S_\varepsilon \sigma)} \frac{a}{\varepsilon} \qquad \textbf{(eq 3.4) [3.1]}$$

While $(1-a)$ is the solar reflectance, $(1-\varepsilon)$ is thermal reflectance. To obtain low values of a **and** ε, a reflecting metal such as aluminium is required, deposited by vacuum evaporation. Evaporated aluminium film solar absorbance is around 0.08 but thermal emittance at 27°C is lower, around 0.01. Aluminium has a high $\frac{a}{\varepsilon}$ ratio unsuitable for external satellite coatings, which should maintain low temperatures to avoid overheating. Lower $\frac{a}{\varepsilon}$ values are obtained by coating aluminium with high emissivity (low reflectance) materials that don't absorb much at short wavelengths. Silicon and aluminium oxides, SiO_2 and Al_2O_3 respectively, are suitable. Thermal absorbance and emittance increase with deposited layer thickness.

Solar absorbance changes little, but we can obtain desired $\frac{a}{\varepsilon}$ ratios between 8, with pure Al, to under 0.2 by controlling SiO_2 thickness. A 1 micron SiO_2 layer, which looks brown to the eye, produces $a = 0.11$ and $\varepsilon = 0.22$, so $\frac{a}{\varepsilon} = 2$, while 4 microns of SiO_2 raises emittance to 0.7 ($\frac{a}{\varepsilon} = 6.36$).

Sensitive remote sensing devices measure radiant temperature. It is important to contrast Earth observation targets from surroundings for imaging and determine surface characteristics to detect specific minerals and quantities such as Sea Surface Temperature (SST). SST depends upon: emissivity, kinetic temperature, thermal properties and heating rate. Sources of surface temperature gain are the absorbed short wave energy (emitted from the sun), balanced against long wave emitted energy from Earth (heat loss) supplemented by anthropogenic heating (urban and power plants) and geothermal sources (volcanoes and hot springs).

Emissivity is essential if **kinetic** temperature (**actual** temperature recorded with a thermometer) is to be *estimated* from T_{rad} recorded by remote sensors.

From Kirchoff's thermal radiation law we find $1 = \rho + \varepsilon$ with ρ = reflectance and ε = emittance, arising from $1 = \rho + a + \tau$ from equations already examined. With hot emissive objects at thermal wavelengths strictly:

$$1 = \rho + a + \varepsilon + \tau \qquad \textbf{(eq 3.5)}$$

However, the *net* result of little absorption and much bigger overall emission is reasonably termed into a single net **ε** factor. For a typical medium, such as water, no heat is transmitted through the medium, it is absorbed and strongly scattered, so **τ** may be considered zero, leaving: $1 = \rho + \varepsilon$, thus:

$$\varepsilon = 1 - \rho \qquad \textbf{(eq 3.6)}$$

The Stefan-Boltzmann black body law is:

$$M_{BB} = \sigma T_{rad}^4 \qquad \textbf{(eq 3.7)}$$

and radiant temperature is a combination of emission and reflection.

$$M_{BB} = \varepsilon \sigma T{kin}^4 + = \rho \sigma T{sky}^4 \qquad \textbf{(eq 3.8)}$$

so

$$T_{rad} = \sqrt[4]{\varepsilon T_{kin}^4 + \rho T_{sky}^4}$$ (**eq 3.9**)

There are several important thermal quantities:

Heat capacity (c) – this is a material's ability to absorb heat. Some, such as water, have high specific heat capacity, meaning they absorb a lot of heat before warming. Water's high heat capacity helps regulate the rate at which air changes temperature, which is why temperature changes between seasons are gradual rather than sudden, especially near oceans. Water has a specific heat capacity of 4.186 joules/gram °C. Other materials have low thermal inertia – for example, dry sandy soils respond quickly to temperature change, reaching high temperature by day and dropping much lower at night. Saudi Arabian and Kuwaiti deserts have large temperature extremes, diurnal variations, day to night.

Thermal conductivity (k) $Wm^{-1}K^{-1}$ is a measure of the rate of heat transfer through a substance; diamond = $1000 Wm^{-1}K^{-1}$ while wood = $0.04 - 0.12 Wm^{-1}K^{-1}$.

Thermal inertia (P) is a measure of material thermal response to temperature changes, units $Jm^{-2}K^{-1}s^{-1/2}$. Thermal inertia is calculated as the square root of the product of density (kgm^{-3}), thermal conductivity ($Wm^{-1}K^{-1}$) and material specific heat capacity ($Jkg^{-1}K^{-1}$). Thermal inertia measures material response to temperature change, increasing with increasing material conductivity, capacity and density. In general, materials with high thermal inertia have uniform surface temperatures day and night. Material properties are measured directly or in situ where:

$$P = (k.\rho.c)^{1/2}$$ (**eq 3.10**)

Thermal inertia *cannot* be determined from remote measurements alone. For a range of rocks, thermal inertia increases with density. Peridotite stores the most heat and pumice the least. But it is important in Earth observation to establish inertia, through computation of Apparent Thermal Inertia, and this is achieved in three stages:

1. Use visible imagery to estimate albedo, A.

2. Use calibrated thermal imagery to estimate radiant temperatures from day and night images.

3. Compute the Apparent Thermal Inertia (ATI):

$$\text{ATI} = \frac{1-A}{\Delta T} \qquad \textbf{(eq 3.11)},$$

where A is the material albedo and ΔT the temperature change between day and night observations.

Example 3.2: Consider sandstone and siltstone:

Material	Day	Night	ΔT
Sandstone	21	10	11
Siltstone	25	8	17

If A = 0.1 and the temperature change 10K, what is the ATI?

$$\text{ATI} = \frac{1-0.1}{10} = 0.09 K^{-1}$$

Materials with high thermal inertia, such as wet clays, resist change and have small temperature ranges. However, shortly after sunset and until two hours after sunrise, objects have similar temperatures regardless of inertia, figure 3.1. Because of differences in *emissivity* and *temperature,* surface differences can generally be distinguished at most other times; for further details, see Sabins [3.2]. The most important variation in thermal remote sensing is the daily solar heating cycle. Figure 3.1 shows typical surface temperature variations over time.

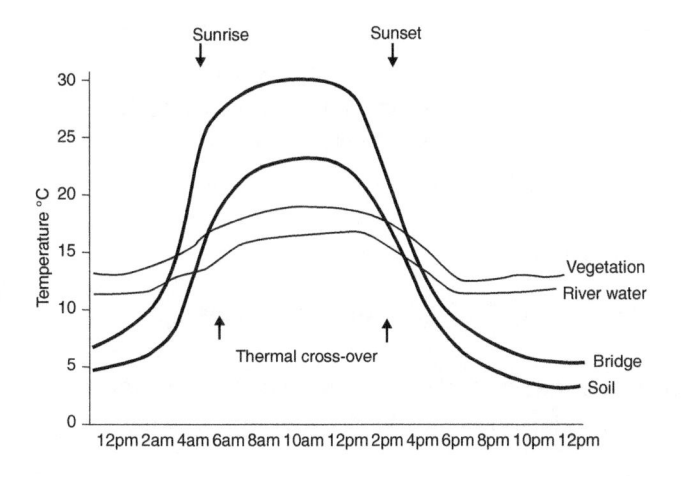

Figure 3.1: *Temperature changes for different environmental surfaces.*

3.3 Spatial variability

This is the biggest problem, as small emissivity changes result in large T_{rad} changes. Most thermal systems have limited pixels to cover a scene (typically 88,000 pixels), from close handheld systems, police and maritime surveillance often helicopter-mounted, to poor resolution satellite systems (one Landsat pixel covering 120 × 120km²). Most natural objects, except water, are selective or 'coloured radiators' with wavelength-dependent emissivity, creating complex Earth observation problems if wide bands are sensed. However, it is usual to use two bands: 3–5 and 8–14 microns; within these, most objects have stable emissivities.

3.4 Principal wavebands

Two key wavebands are used for thermal remote sensing, separated by atmospheric water vapour absorption 5–6 microns: these are the 3–5 and 8–14 micron bands. The 3–5 micron band is useful in volcano studies and night-time oceanography [3.3]. The 8–14 micron band contains the peak of black body radiation spectrum 210–360K (–70 to 90°C) covering room temperature applications. Thermal applications are classed into those governed primarily by man-made heat sources or solar radiation. In the first case, airborne platforms can determine building and structure heat losses. This type of observation is best performed just before dawn, so solar heating effects have the most time to decay away.

3.5 Kinetic temperature

T_{kin} is the temperature recorded by thermometers or other **contact** sensor. This is related to the remotely sensed radiant temperature, **T_{rad}**, because hot objects with a high T_{kin} c emit strongly with high T_{rad}. Object T_{kin} won't equal T_{rad} recorded by a sensor unless it is a black body, hence the importance of knowing surface emissivity, as real objects emit *less* heat than black bodies. For a grey body, T_{rad} is related to an object's emissivity with the formula:

$$T_{rad} = \varepsilon^{1/4} T_{kin} \qquad \textbf{(eq 3.12)}$$

If emissivity is **not** accounted for, recorded T_{rad} data **underestimates** T_{kin}.

Current emissivity estimates are the primary limitation in remotely sensed thermal infrared data accuracy. Emissivity is required ±0.02 before T_{kin} is estimated to ±1°C.

> **Example 3.3**: Consider the difference in radiant temperature for a kinetic temperature of 300K and objects with emissivities of 1.0 and 0.4.
> Using the equation $T_{rad} = \varepsilon^{1/4} T_{kin}$ in the first case:
>
> $T_{rad} = (1.0)^{1/4}(300) = 300K$ **(eq 3.13)**
>
> while in the second:
> $T_{rad} = (0.4)^{1/4}(300) = 238.6\ K$, an underestimated difference of 61.4K!

Even if emissivity is accounted for, it is only a *predicted* T_{kin} from which T_{rad} and ground truthing are required to find T_{kin}. T_{rad} is the detected radiation from the surface, while T_{kin} is the *actual internal* object temperature, figure 3.2. T_{rad} recorded by a radiometer predicts T_{kin} for freshly dug soil. On a summer morning, soil surface is moist with evaporative heat loss, predicting T_{kin} several degrees *cooler* than the actual T_{kin}, recorded a few centimetres beneath the surface. By early afternoon, soil surfaces have dried out, with predicted temperatures T_{kin} cooler than actual T_{kin}.

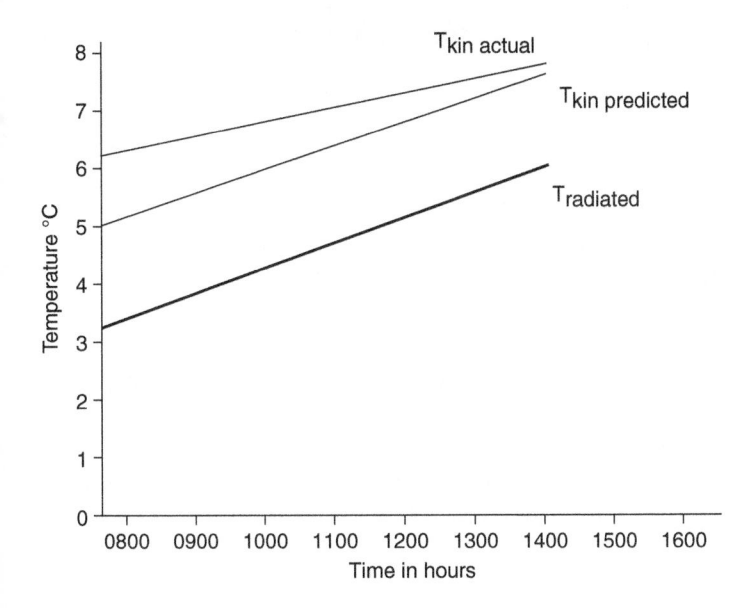

Figure 3.2: *Kinetic vs radiant vs predicted temperature.*

Remote sensing methods estimate temperature differences. Measuring actual temperature is difficult, but with 'ground truthing', calibration and multiple spectral wavelength brightness calculations, accurate temperatures are achieved by satellite remote observation methods.

3.6 Thermal crossover

For good imaging contrast, radiation reaching an imager from two surfaces must be different. We require sufficient contrast to separate them. For radiated intensity given by $\varepsilon\sigma T^4$ for two different surfaces '1' and '2' to be distinguished, $\varepsilon_1 \sigma T_1^4$ must differ from $\varepsilon_2 \sigma T_2^4$. The condition where they are the same is *thermal crossover*:

$$\varepsilon_1 \sigma T_1^4 = \varepsilon_2 \sigma T_2^4 \quad \textbf{(eq 3.14)}$$

Example 3.4: Consider a bridge over a river in northern hemisphere winter. Both surfaces are indistinguishable from each other as determined by a UAV taking IR images at dusk. If the river was at 2°C and the emissivities of the river and bridge 0.7 and 0.95 respectively, the bridge temperature can be found by considering thermal crossover where: $\varepsilon_B \sigma T_B^4 = \varepsilon_R \sigma T_R^4$

Rearranging: $T_B^4 = (\varepsilon_R/\varepsilon_B) T_R^4 \quad T_B = (\varepsilon_R/\varepsilon_B)^{1/4} T_R$

Find 4th root and substituting:

$T_B = (0.7/0.95)^{1/4} (273 + 2) = 254.8K = -18.2°C$

3.7 Heating rate

Ground heating is mostly the result of daily solar intensity and insolation absorption rate. The solar heating cycle reaching ground is reduced by obstructions such as trees, clouds and buildings and affected by aspect and slope.

3.8 Interaction of thermal infrared wavelengths of electromagnetic radiation with water (the hydrosphere)

Internal convection maintains constant water surface temperatures in large water bodies, unlike soil, which swings in large heating/cooling cycles. Surface waters of different salinity and temperature can be viewed with thermal imagery.

3.8.1 Sea Surface Temperature

Sea Surface Temperature (SST) is of great oceanographic and meteorological importance – for example, El Niño. SST is deduced from calibrated IR data, as pure water emissivity = 0.993. However, IR signals are characteristic of water surfaces to only the top 100 microns depth and may not indicate water **internal** bulk temperature. Although bulk water temperatures may be warmer than surfaces, thermal sensors record **only** surface temperature, and records T_{rad} less than actual T_{kin}.

From a satellite perspective, atmospheric attenuation reduces intensity and is modified by sky reflections. Individual band satellite observations may deviate from calibration by 10K, not much use when trying to make accurate climate change assessments! Incident flux calculations are complex, including contributions from direct solar and sky radiation, geothermal heat and surface re-radiation. It depends on whether skies are clear or overcast, geographical latitude, time of year and surface emissivity. Various calculation methods exist; however, the principle idea uses two different IR spectral bands with linear relationship of the form:

SST(True) = $AT_1 + BT_2 + C$, where T_1 and T_2 are 'brightness' temperatures measured at two wavelengths, for example, 8 and 10 microns, and A, B and C are constants.

Example 3.5: If T_1 = 320K, A = 0.91, B = 0.14, SST(True) = 325K, and C = −16, what is the brightness temperature T_2 at 10 microns?

Using SST(True) = $AT_1 + BT_2 + C$

T_2 = (SST(True)-AT_1-C)/B = (325 − 0.91 × 320 + 16) = 45K

3.9 Interaction of thermal infrared wavelengths of electromagnetic radiation with vegetation and chlorophyll

Vegetation canopy thermal properties are more complex than water, soil or urban because vegetation absorbs more solar energy in the visible, which it re-emits in the infrared to maintain overall energy balance. Diurnal (daily) canopy range is small so a canopy will be at a different temperature to surrounding air most of the day. On warm UK summer days, leaf temperature can be 15° below air temperature at midday, yet 5°C above air at midnight. During warming and cooling periods, leaf temperatures follow a cycle dependent upon vegetation thermal inertia. Vegetation canopy temperature is modified by large leaf area and high transpiration rates. Large canopy areas result in high radiation rates, sometimes 30 times more than surrounding soil, reducing canopy temperature significantly.

Transpiration is the main mechanism by which plants lose water through evaporation; it keeps them cool during hot days and is controlled by specific vegetation characteristics and environment. Short-term environmental factors – atmospheric humidity, light availability, temperature, wind speed and soil moisture – affect this. On sunny, hot, windy days with moist soils and dry air, transpiration rates are high and canopies cooler than surrounding air as leaf water moisture is driven from the surface, which cools as 'work' is done. ASTER composite images, visible and NIR, show vegetation especially clearly.

3.9.1 Vegetation moisture content

Leaf moisture content determines leaf surface emissivity. Dry leaves have emissivity = 0.96, while moist leaves have emissivity closer to 0.99. Canopy moisture content varies seasonally, spatially and diurnally. Changing the volumetric soil water content 0.047 to 0.44cm^{-3} per cubic centimetre (half) reduces maximum daily maximum-minimum surface soil temperature from 40 to 20°C.

3.9.2 Sensor angle

Vegetation canopy T_{rad} is related to sensor angle and emissivity. At midday, a vertical view of soil and vegetation records high temperatures, while oblique views see more vegetation and record only canopy temperature T_{rad}, appearing lower [3.4].

3.10 Interaction of thermal infrared wavelengths of electromagnetic radiation with snow and ice

Mountain glaciers have been investigated using thermal imaging. Under incoming direct short-wave radiation, glacier debris generally has stronger emission than ice due to higher temperatures, leading to strong thermal signals [3.5]. Taschner and Ranzi [3.6] showed emission differences exist between debris-covered glaciers and surroundings, and how differences facilitate glacier outline delineation. During daytime, long-wave radiation emitted from terrain depends on incoming short-wave radiation [3.7]. Hence night-time or early morning imagery better indicates ground thermal differences.

Much of Earth's surface is covered by frost, snow and ice. Their spectral emissivities are a significant input to the global radiation balance and measurement of reflectance spectra helps provide emissivities using Kirchoff's law. Spectra show that snow emissivity departs significantly from black body behaviour in the 8–14 micron region. Snow emissivity decreases with increasing particle size and increasing density due to packing, while emissivity increases due to meltwater [3.8]. Furthermore, snow and ice spectral emissivity in the 8–14 micron atmospheric window is a fundamental property for determining snow surface temperatures from space [3.9] and detecting clouds over cold snow surfaces in the polar night [3.10].

Earth's polar regions are considered by seafarers to be the most challenging and treacherous environments. Thermal imaging helps seafarers find safer passage through ice. Glacier ice is difficult to track with marine radars as echoes are scattered by air bubbles and ice imperfections. Even radar echoes from large icebergs are less than ship targets because of reduced ice reflectivity when compared with steel. Ice detection is difficult, especially if it has a smooth profile. Small pieces are hard to detect and easily lost among 'sea clutter'. During daylight hours, radar's inability to detect ice can be compensated by visual lookouts, but this requires good visibility. However, in dark polar nights this task becomes difficult and further restricted by snow or fog, common in Arctic open waters during winter. Thermal imaging cameras work well, with a wide camera range available for various resolutions to meet maritime needs.

3.11 Interaction of thermal infrared wavelengths of electromagnetic radiation with soil

Soil T_{rad} is determined by moisture content, which determines emissivity. Wet soil is cooler during daytime and warmer at night. The depth soil moisture ceases to affect changing T_{rad} values and varies with soil type from mm to cm. Sandy soils heat quickly in daytime and cool rapidly at night. The method of estimating Soil Thermal Inertia (STI) primarily follows that discussed for ATI using midday heat flux, and the time between daily maximum and minimum surface temperatures. The difference between STI and ATI means it is easier to get midday soil flux from remote data. Typical STI is found to be around $1370.6 Jm^{-2}K^{-1}s^{-1/2}$ in the growing season and $1061.2 Jm^{-2}K^{-1}s^{-1/2}$ in the non-growing season [3.11]. Rainfall and frequent irrigation is the primary reason for higher growing season STI because high soil water content usually produces high soil thermal inertia.

3.11.1 Soil temperature

Vegetation canopy T_{rad} is a mixture of soil and leaf temperatures. At midday, T_{rad} is a mixture of high soil temperature (soil visible with low transpiration) and lower leaf temperature (high transpiration with strong cooling effect). Canopy T_{rad} is higher than leaf T_{kin}. To determine leaf T_{kin}, relative area and soil temperature are needed; see section 3.9.

3.12 Interaction of thermal infrared wavelengths of electromagnetic radiation with rocks and minerals

Satellite thermal remote sensing images have been used for mineral exploration since Landsat. This application relies on a sensor's capability to register unique spectral signatures with specific mineral deposits. Gold is one important mineral searched with satellite remote sensing imagery over the last 30 years. The methodology used is indirect – gold is not sensed directly; however, minerals formed in *association* with gold are detected based on their spectra. Clay minerals occurring in alteration zones are associated with gold, with spectral signatures mostly in the short-wave infrared. They can locate sites where gold deposits are *likely* to occur, saving the mining industry time and money in exploration. A promising satellite survey is supplemented by a higher resolution aerial survey like TASI. There is little contrast between water thermal inertia and typical minerals, but they can be distinguished, as water is normally cooler than rock during daytime because of evaporation. At night, water surfaces are warmer, dry vegetation is distinguished from bare ground and rock, as vegetation and ground are warmer than rock, while vegetation provides insulation. Canopies scatter heat back to ground, often noticed on cold nights with 'rings' forming round frost-free trees. Common Cypriot rock types have conductivities between 0.4 and 44.2Wm^{-1}K^{-1}, diffusivity values 0.3–1.9 × 10^{-6}m^{2}s^{-1} and specific heat capacities 0.5–1.5JK^{-1}kg^{-1}[3.12]. Most rock and mineral inertias increase with density. Sandstone stores the most heat and rhyolite the least.

3.13 Satellite thermal IR systems

Thermal images have been acquired for decades, starting with a range of environmental and Met unmanned satellite programs, including Landsat TM, AVHRR and HCMM. Thermal inertia mapping was first carried out from space with HCMM (Heat Capacity Mapping Mission), carrying a thermal radiometer (10.5–12.5 microns), the HRIR, with 0.4K accuracy. Distinguishing rock types is difficult, but ASTER composite SWIR bands (4, 6 and 8) show differences in clays, carbonates and sulphate minerals, while composite thermal bands (10, 12–13) show quartz, carbonates and volcanic rocks with different false colours.

Thermal inertia mapping is used in archaeological surveys. If one material is buried within another and the materials have different thermal properties, heat flow distorts and generates surface anomalies. Thermal variation extremes and surface material heating rates are determined by material thermal conductivity, capacity and inertia. The EOS/Terra platform, launched in 1999, carried the Advanced Space-borne Thermal Emission and Reflection (ASTER) radiometer. ASTER is a multispectral imaging radiometer covering VNIR, SWIR and TIR wavelengths with 14 bands, and 15–90m spatial resolution, designed to improve understanding of near Earth surface processes.

3.13.1 Space-borne thermal imagers

The first satellite specifically designed for Earth observation was the Earth Resources Technology Satellite-1 (ERTS-1), renamed **Landsat**. The US Geological Survey required information about geological structures, while the Department of Agriculture needed information for managing and monitoring agriculture. The mission was a success, producing images for both and for other applications. Several Landsat satellites have operated since, with Landsat 8 the latest and Landsat 9 planned for launch in 2020. The first three, Landsat 1–3, carried two main sensors: Return Beam Vidicon (RBV), a video-imaging instrument, and MultiSpectral Scanner (MSS).

The short wavelength bands are shown (figure 3.3). Earth peak radiated energy is 9.7µm, and TM band 6 records 10.5–12.5µm radiation, avoiding strong ozone absorption 9–10µm. Landsat 4 and 5 were placed in near-polar, sun-synchronous orbits at 705km. Both carried an MSS and another multispectral scanner, the Thematic Mapper (TM). Landsat 5 operated until 2000. There are several common thermal imagers: Landsat – one channel, AVHRR – two channels, ATSR – three channels, MODIS and ASTER – several channels. Thermal detectors should be cooled as temperature changes create noise, with deep space acting as a calibration target.

3.13.2 Thematic Mapper (TM)

TM is an optical sensor detecting reflected energy and emitted heat from Earth. It has a line scanner with a mirror rotating across an 18.5km swath width of ground, as the satellite moves along its orbit. The sensor array collects data on scans and the satellite's forward motion allows successive surface strips to be covered. The system is designed so mirror movement fits in with platform movement, each strip next to the previous one.

Principles of Earth Observation

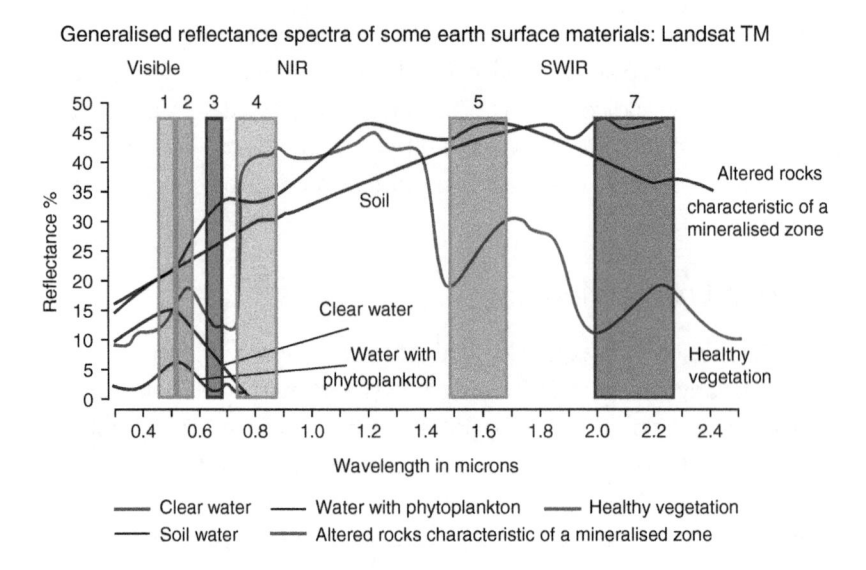

Generalised reflectance spectra of some earth surface materials: Landsat TM

Figure 3.3: *Common surface reflectance spectra within Landsat TM bands.*

Energy from Earth's surface strikes a mirror, passes through optical lenses and on to detectors. Before it reaches the detectors, it is split into seven bands. Photoelectric detectors are used for six but heat detectors record thermal heat with 120m IR band 6-pixel resolution.

Channel waveband (micrometres)	Spatial resolution (m)
1 0.45–0.52	30
2 0.52–0.60	30
3 0.63–0.69	30
4 0.75–0.90	30
5 1.55–1.75	30
6 10.40–12.5	**120**
7 2.08–2.35	30

Table 3.2: *Landsat 4 and 5 TM waveband vs spatial resolution.*

3.13.3 Advanced Very High Resolution Radiometer (AVHRR)

The AVHRR, carried on NOAA polar-orbiting satellites, recorded one visible, one reflected NIR and three IR bands. AVHRR spectral range (figure 3.4) band 3 (3.5 microns) detects hot targets, monitoring active volcanoes and forest fires.

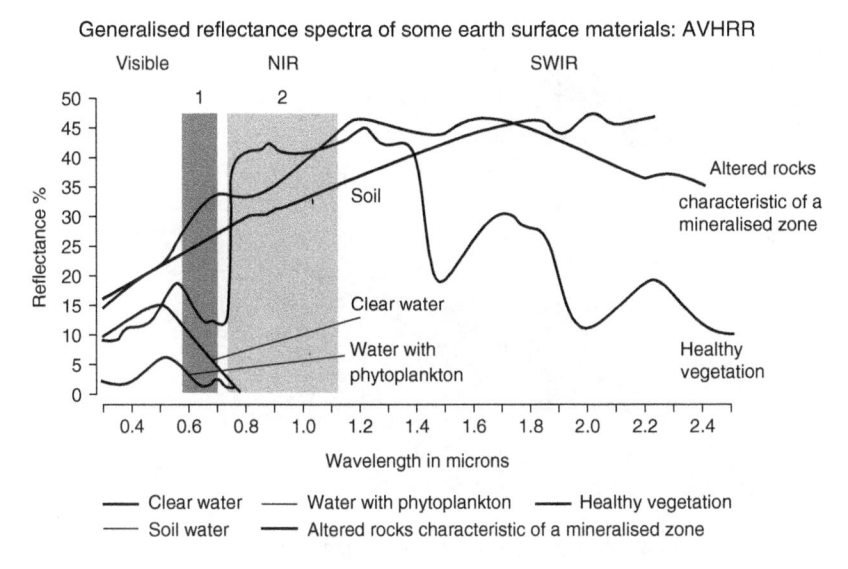

Figure 3.4: *Common surface reflectance spectra within AVHRR bands.*

3.13.4 Heat Capacity Mapping Mission (HCMM)

NASA launched HCMM to record day and night images covering 700 × 700km, 1978–80.

3.14 Thermal IR spectra

Thermal spectrometers record spectra emitted from materials in the 8–14µm band. To obtain images of spectral information, the Thermal IR Multispectral Scanner (TIMS) was developed, recording six bands: 8.2–8.6, 8.6–9.0, 9.0–9.4, 9.4–10.2, 10.2–11.2 and 11.2–12.2µm. TIMS has IFOV = 2.5mrad and FOV = 80°.

3.15 Non-imaging systems

Non-imaging systems, including optical, are systems concerned with optimal radiation transfer between source and target. Unlike traditional optics, techniques don't attempt to form images; instead, an optimised system for radiative transfer from source to target is desired.

3.15.1 Infrared radiometer

The basic radiant temperature sensor is the radiometer, a non-imaging device recording radiant temperature within its IFOV, containing collection optics, filters limiting spectral range, electronics to amplify signals and a display.

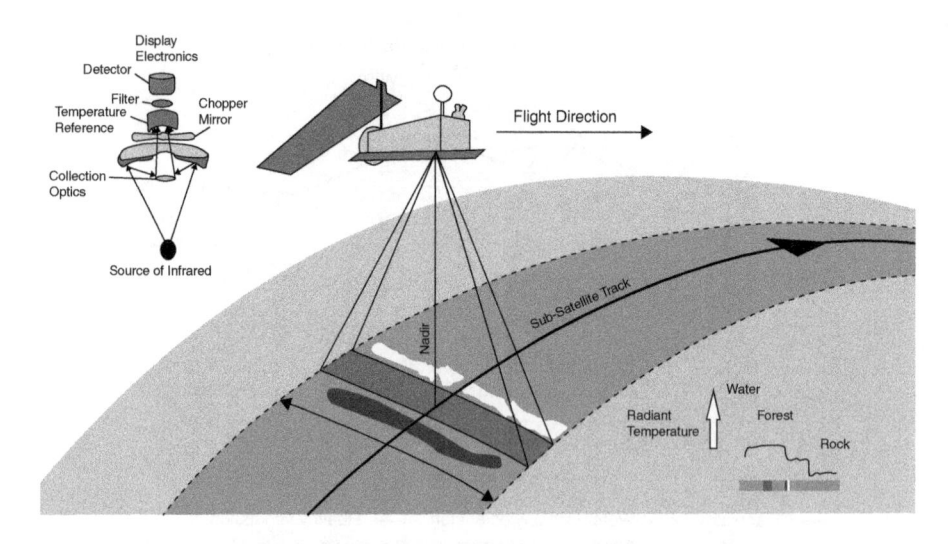

Figure 3.5: *Schematic of a radiometer and overflight of water, forest and rock terrain.*

The instruments collecting optics focus ground-radiated energy on to the detector. Mirrors are used, as glass absorbs infrared. A rotating chopped mirror provides an alternating view of ground and a calibrated internal temperature reference. A radiometer is positioned vertically to view ground directly below (at *nadir*) along the flight direction, including Nimbus 5's THIR, used to assess absorptive moisture in the water vapour absorption band, and the Surface Composition Mapping Radiometer (SCMR) on Nimbus at 8.4–9.5 and 10.2–11.4 microns, figure 3.5.

At any instant, the radiometer senses radiation within its IFOVβ, determined by the optical system and detector element size. All radiation propagating towards the instrument within the IFOV contributes to detector response. The ground surface measured within the IFOV corresponds to arc length = $r\beta$, written as:

D = Hβ   **(eq 3.15)**,

where D = the viewed circular ground area diameter, H = height above terrain and β = IFOV of the system in radians. Ground area sensed is termed the system's *spatial resolution*.

A radiometer with $IFOV_\beta$ = 2.5mrad, operated 100km above terrain, has a ground resolution element (or cell diameter) of 250m. As the sensor integrates measurements over the IFOV, objects with different temperatures won't be differentiated within the cell. A small IFOV is desired for high spatial detail in any band, analogous to resolution problems in Chapter 5. A thermal satellite-based radiometer operates in the 8–14μm band, ideally with equal response to a wide infrared band.

3.15.2 Spectrometer

A spectrometer selects a *desired* wavelength or IR band, achieved by dispersing incoming radiation with prisms, gratings, mirrors or filters. Scanning systems are classed as thermal or photon detectors. Airbus's Defence and Space UV-VIS-NIR-SWIR push-broom grating spectrometer TROPOMI, used on the Sentinel-5 Precursor mission, has five channels, providing 7 × 7km ground resolution, *spectral range* 270–495, 710–775, 2305–2385nm, with spectral *resolution* 0.25–0.55nm. Earth resource scanners use photon detectors because of their sensitivity. Several photon detectors from 0.4 to 14μm are given, table 3.3.

Cooling temperature (K)	300	195	77	35
1.5–1.8 microns	InAs InSb HgCdTe	InAs HgCdTe	InAs	
2–2.5 microns	InAs InSb HgCdTe	InAs InSb HgCdTe	InAs InSb HgCdTe	
3.5–4.1 microns	HgCdTe InSb	HgCdTe InSb	InSb HgCdTe	
8–12 microns			HgCdTe PbSnTe PbSbTe HgCdTe	GeHg

Table 3.3: *Common photodetectors.*

Detector cooling is generally needed except for silicon 0.4–1.1 and germanium 1.1–1.75μm. It is important to provide in-built calibration, using tungsten lamps or controlled black body sources. Maximum radiance resolution is limited by detector cooling; liquid nitrogen cooling 77–100K provides resolution in the 8–13μm range. Satellite MSS use solid cryogen closed-cycle refrigerators or passive radiation to space methods. With solid cryogens, a pre-cooled solid gas is stored in a high-performance dewar and vented into space. Longer lifetime closed-cycle refrigerators provide higher cooling capacity/weight/size ratios than solid cryogens but require higher power. Passive cooling is attractive because of reduced power and longer life; however, cooling capacity is generally too low for large arrays. Simple handheld radiometers are used from aircraft and helicopters – for example, SASI 1000A Airborne Hyperspectral SWIR imager (0.95–2.45 microns) with 20 spectral channels, 40°FOV and 1,000 spatial imaging pixels. Many systems are available [3.13]. Space-based radiometers include AMSR-E, AVHRR, Aquarius, ICESat and MODIS.

3.16 Far infrared thermal imaging sensors

Infrared line scanners and thermal imaging systems include commercial models providing accurate, real-time imaging for industrial applications.

3.16.1 Thermal scanners

Thermal imaging 'views' vessels by their emitted heat. Older cameras used an objective lens to focus a scene on to a cooled detector array via a rotating mirror. Thermal scanners build a 2D profile of data in **swaths** beneath the aircraft or satellite. The simplest imager carries a single detector scanning multiple lines or narrow strips at 90° to the flight direction in the same way as the TM (see 3.13.2). Images are constructed, having more detectors to increase scanning update rate such as the InfraRed Line Scanner **IRLS**. Some systems allow a sensor array to sweep an area, and conical scanning can be used. IRLS systems are used extensively from aircraft and helicopters, with temperature resolution of 0.25°C at 0°C. Surface temperature is detected to accuracy below 1°C. Signals are processed to form thermal maps providing temperature variations. Figure 3.6 (see plate section) shows a Landsat 8 NASA Thermal InfraRed Sensor (TIRS) image.

Thermal imagers on satellites, such as THEMIS on Mars Odyssey, image visible and thermal bands. A thermal camera is a camera core composed of electronics, IR lenses and a detector (Uncooled Focal Plane Array – UFPA) made of pixels (each a sensor in its own right). The 2016 commercial market saw the sale of some 1.2 million uncooled thermal cameras from companies such as Raytheon, Thales, Northrop Grumman and Elbit Systems. Satellite IRLS systems generate data indicating *relative* thermal energy differences. To achieve *absolute* temperature relationships from scanner data, extra constraints are met, which are not discussed here. A high-quality display is obtained using a detector array of elements matched to a suitable scanning system as a halfway house to a focal plane array. In a banded system, scans are generated by a rotating drum with progressively angled facets, see figure 3.7 (left). Scanning sensors are essential for geostationary satellites without relative motion. Scanning rate provides a compromise between sensitivity, resolution and scanning time – for example, HCMM and Coastal Zone Color Scanner (CZCS) radiometers are space-based systems.

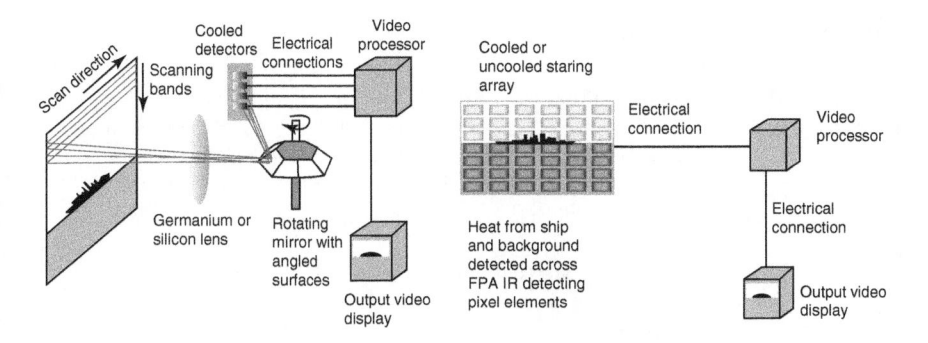

Figure 3.7: (left) *Scanning imager;* (right) *Staring focal plane array.*

Today, most radiometers use FPA imagers as they have few or no moving parts, see figure 3.7 (right). 'Staring arrays' form 2D pixel arrays at the lens focal plane (figure 3.7) and are used for non-imaging applications, that is, spectrometry and lidar. Staring arrays are distinct from scanning imagers as they image the entire FOV at the same time without scanning, typically 320 × 256 pixels. Arrays include Indium Antimonide (InSb), Quantum Well Infrared Photodetectors (QWIP) and Indium Gallium Arsenide InGaAs detectors, affordable even for non-military users such as schools. The commonest FPA types are InSb, InGaAs, HgCdTe (MCT) and Quantum Well devices. The latest use low-cost, uncooled micro-bolometers.

3.17 Detectors for thermal infrared radiation

IR detectors are suitable for different wavelengths, mentioned briefly here. Extensive detector discussion is provided elsewhere, for theoretical radiometry, practical issues such as detector noise, figures of merit and the modulation transfer function, with examples and questions [8.14]. InGaAs PIN photodiodes are appropriate for the 0.9–2.6μm region (figure 3.8), but cannot detect longer wavelengths. Thermopiles, InSb photoconductive detectors and photovoltaic detectors cover 1–11μm (figure 3.8), sensitive 1–25μm, see table 3.4.

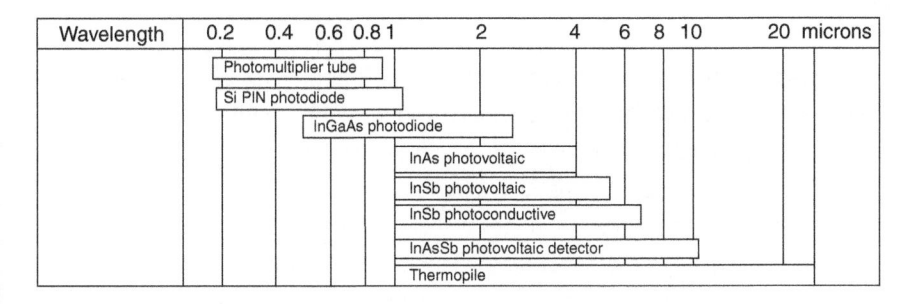

Figure 3.8: *Typical thermal detectors.*

Detector type	Photovoltaic detector	Photoconductive detector	Thermopile
Spectral response	Typically narrow	Typically narrow	Wide
Detection method	IR photons interact with semiconductor charge carriers, generating current	IR photons change semiconductor material; resistance measured as output voltage	IR photons cause temperature changes, converted into output voltage
Sensitivity	High	Medium	Low
Response time	Typically 1ms or less	Microseconds	Milliseconds
Cooling	Can be cooled or uncooled	Can be cooled or uncooled	Uncooled
Cost	Most expensive	In between photovoltaic and thermopile costs	Low cost

Table 3.4: *Typical IR detector characteristics.*

Key IR detectors are *Mercury Cadmium Telluride* (MCT), which operates up to 15μm. MCT or $Hg_{1-x}Cd_xTe$ response can be varied to change cut-off wavelength between 3 and 5μm. Typical pixel size is 45μm². *Bolometers* contain materials whose resistance varies with temperature expressed as

$$R = R_0 e^{\frac{\beta}{T}} \quad \textbf{(eq 3.16)}.$$

R_0 is ambient resistance at nominal temperature, β the material constant.

Thermopiles are a series of thermocouples, each composed of different materials connected in series, forming a cold reference junction and hot measuring junction. Absorbing infrared causes the hot junction to warm, resulting in a temperature difference between junctions. Typical thermopiles are sensitive to 3–5μm. A potential difference is generated across the junctions between dissimilar metals junctions held at different temperatures [3.15]. Like the bolometer, the thermopile has a long response time, around 10ms, and isn't sensitive, but responds up to 30 microns.

Pyroelectric detectors have crystals that undergo redistribution of internal charge to temperature change on the crystal surface and result in a potential difference, achieved by 'chopping' incident radiation at 1kHz, and measuring output, which alternates at this frequency. Alternatively, changes are achieved by 'panning' cameras across a scene to create changes without chopping. Older naval firefighting TICs used pyroelectric detectors operated in 'chopped' or 'pan' modes. The pyroelectric coefficient p measures the electric polarisation rate of change with respect to temperature. In conventional pyroelectrics, internal polarisation P is equivalent to electric displacement D with $p = \frac{dP}{dT} \approx \frac{dD}{dT}$ when the applied electric field E = 0.

Scanning sensors are essential for geostationary satellites without relative motion. Scanning rate is controlled, providing a compromise between sensitivity, resolution and scanning time. HCMM and CSCS radiometers are space-based system examples.

3.18 Advanced cooled FPAs

Platinum silicide (PtSi) is a semi-conductor material with 5.7μm cut-off and 4.3μm maximum response. Indium Antimonide (InSb) operates in the 3–5μm band with 4.7–4.8μm peak response. InSb is more sensitive than PtSi. The Aladdin US Navy and Astronomy InSb array has 27μm pixels in four 512 × 512 quadrant arrays.

Quantum Well Infrared Photodetectors (QWIP) use GaAs semiconductor technology. A typical GaAs/Al$_x$Ga$_{1-x}$As system is engineered for peak spectral response 3–19μm with 0.4–0.5μm bandwidth. JET and Amber designed a cooled LWIR camera, combining JET's success with QWIP's and Amber's 256 × 256 FPA to monitor forest fires from space.

IR photodetectors made from laser fabrication materials promise a new generation of cameras. Quantum wells are used in lasers to generate light from current, but IR cameras reverse this process. In conventional QWIP devices, electrons exist in two energy bands, a low energy 'valence' band or a high-energy 'conduction' band. Photons striking conventional surfaces liberate electrons into the conduction band, generating an electrical signal if voltage is applied across the detector. Recent QWIPs are *so* sensitive, photons are detected if they have enough energy to shift electrons *within* bands. GaAs/InGaAs materials are fabricated into large arrays sensitive to 3–5 and 8–13μm 'windows'. As Earth's atmosphere is transparent across these regions, QWIP cameras mounted on LEO satellites can record higher resolution images, offering improved detector uniformity and lower unit cost, simple fabrication and higher yield.

Future developments include Dual Band sensors offering better performance with MIR and FIR sensors linked together relatively cheaply through **image fusion**, combining different image types into a single 'hybrid' display, which is more informative than individual pictures. Scientists have merged radar and optical satellite data, such as **ERS-1**, to study agriculture. Cranfield University UK is investigating fused NIR (0.75–1.1μm), with good contrast and spatial resolution but limited range, and FIR (8–12μm). Merging images and tracking key features 'edge tracking' and neural networks improve vision.

3.19 Emerging Uncooled FPA (UFPA)

Several devices – ferroelectrics, microbolometers and pyroelectrics – have one common characteristic: *no cooling* is required for temperature stabilisation, ideal for low-cost devices. Texas Instruments introduced the first ferroelectric UFPA. There is interest in microbolometers because costs may be reduced below ferroelectrics. LWIR uncooled devices are ideal for room temperature applications (8–12μm). Uncooled microbolometers and pyroelectric sensors rely on pixels absorbing IR. A transistor switch beneath each pixel allows them to be actively addressed. UFPAs are ideal for medical thermography with potential for low-cost and long-wave response. A promising future awaits thermal imaging in many disciplines, including satellite remote Earth observation.

ESA's Gaia satellite is the largest FPA ever: 106 CCD imaging sensors operated together with 1,000 million pixels, already providing incredible deep space images. Analysts anticipate micro-optics and multi-hyperspectral systems (imaging a scene and simultaneously determining spectral content). The basic principle of hyperspectral systems relies on the diffraction grating, separating radiation into different bands according to wavelength.

In summary, future systems will make extensive use of multispectral image fusion and have less weight, smaller size, reduced power consumption, higher resolution, wider IFOV and increased range for detection/recognition and identification. Some systems will be discussed later.

Questions

3.1

(i) What is $M_{\lambda \, sun}$ at 0.5 microns?

(ii) What is the incident spectral radiant flux at the top of the Earth's atmosphere?

(iii) If $M_{BB} = \sigma T_{rad}^4$ and radiant temperature measured by a sensor is a combination of an emitted component with emissivity = 0.2 and a reflected component 30%, T_{kin} = 30°C, T_{sky} = −40°C, find T_{rad}.

3.2 $ATI = \dfrac{1-A}{\Delta T}$ where A is the material's visible albedo, where ΔT is the temperature change between day and night observations.

Consider two materials:

Material	Day	Night	ΔT
Material 1		22	8
Material 2		28	6

If A = 0.8, what is the ATI?

3.3 A man is overboard in seas at 9°C, with emissivity = 0.984. The crew member has emissivity = 0.80 and the initial surface temperature = 23.5°C. Could the crew member be detected with a thermal camera if it can discriminate 25 per cent intensity differences? Justify your answer with appropriate calculations. As time passes his temperature falls, becoming harder to detect and then easier. What is happening? At what temperature would he be hardest to detect? Why is it important to find the man well *before* thermal crossover occurs?

3.4 Explain what is meant by emissivity. Consider the difference in radiant temperature for a kinetic temperature of 300K and two objects with emissivities 0.8 and 0.5 respectively. A filament light bulb is heated to a radiant temperature of 6900K. What is its peak radiated wavelength?

3.5 If T_1 = 310K, A = 0.88, B = 0.15, SST (True) = 320K and C = −18, what is the brightness temperature T_2 at 10 microns?

3.6 Considering figure 3.3, explain which two Landsat TM bands will show (a) very good, (b) good and (c) poor discrimination between healthy vegetation and altered rocks. What is the approximate percentage difference between altered rocks and healthy vegetation? What is this value expressed in dB?

3.7 Describe the importance of the Planck black body energy distribution to remote sensing. Make clear the distribution's key features, using figures where appropriate.

What is the theoretical total emissive power M of Earth in Wm^{-2}, at a surface temperature of 300K? At what peak wavelength does this power occur? Use $\sigma = 5.67 \times 10^{-8}$Wm^{-2}K^{-4}.

3.8 Define the following terms and briefly detail their relevance to remote sensing: grey body and radiance.

Which of the following have thermal bands: ETM +, ASTER, IKONOS and TM?

3.9 Discuss how AVHRR and MODIS provide active forest fire detection.

(i) Calculate the wavelength of peak thermal exitance for the following: (a) An Arctic snowfield at −15°C, (b) molten lava at 1160°C and (c) hot desert sand at 48°C.

(ii) Calculate for each surface the radiant exitant intensity using the following emissivity values: Arctic snowfield $\varepsilon = 0.94$, molten lava $\varepsilon = 0.82$ and hot desert sand $\varepsilon = 0.43$. $\sigma = 5.67 \times 10^{-8}$Wm^{-2}K^{-4}.

3.10

(i) Discuss how to distinguish differences between obsidian and a concrete walkway in the thermal bands 8–10 microns. If the concrete temperature is 20°C, at what temperature will the obsidian surface be for thermal crossover to occur? Obsidian $\varepsilon = 0.862$ and concrete walkway $\varepsilon = 0.966$.

(ii) Calculate thermal inertia from the data: thermal conductivity = 1.13W/(m.K), density = 2000kg/m^3 and thermal capacity = 2000J/(kg.K).

References

[3.1] *Thin Films*, KD Leaver and BN Chapman (Wykeham Publications Ltd, London, 1971, ISBN 0 85109 230 6).

[3.2] *Remote Sensing: Principles and Interpretation*, 3rd Edition, Floyd F Sabins (W.H. Freeman and Company, San Francisco, 1996, ISBN 9780716724421).

[3.3] 'Channel Selection for the Next Generation Geostationary Advanced Baseline Imagers', TJ Schmit et al., NOAA/NESDIS, Office of Research and Applications, Advanced Satellite Products Team (ASPT), ams.confex.com/ams/pdfpapers/54285.pdf

[3.4] *Principles of Remote Sensing*, PJ Curran (Longman Group Limited, London, 1985, ISBN 0582300975), p.35.

[3.5] 'Optical constants of ice from the ultraviolet to the microwave: A revised compilation', S Warren and R Brandt, *Journal of Geophysical Research Atmospheres*, Vol. 113(14), 2008 doi: 10.1029/2007JD009744.

[3.6] 'Comparing the opportunities of LANDSAT-TM and ASTER data for monitoring a debris covered glacier in the Italian Alps within the GLIMS project', S Taschner and R Ranzi, (2002), Conference: Geoscience and Remote Sensing Symposium, 2002. IGARSS '02. IEEE International, Vol. 2.

[3.7] 'First results and interpretation of energy-flux measurements over Alpine permafrost', C. Mittaz et al., *Annals of Glaciology*, Vol. 31, 2000, pp.275–280.

[3.8] 'Measurements of thermal infrared spectral reflectance of frost, snow, and ice', JW Salisbury et al., *Journal of Geophysical Research: Solid Earth* Vol. 99(B12), 10 December 1994, pp. 24, 235–24, 240.

[3.9] 'Arctic ice surface temperature retrieval from AVHRR thermal channels', J Key and M Haefliger, *Journal of Geophysical Research: Atmospheres*, Vol. 97(D5), 20 April 1992, pp.5885–5893, 10.1029/92JD00348.

[3.10] 'In-situ measured spectral directional emissivity of snow and ice in the 8–14 um atmospheric window', M. Hori et al., *Remote Sensing of the Environment*, 100(4), 2006, pp.486–502.

[3.11] 'An Empirical Method of Estimating Soil Thermal Inertia', J Tian et al., *Advances in Meteorology*, Vol. 2015, dx.doi.org/10.1155/2015/428525.

[3.12] 'Measurement and analysis of thermal properties of rocks for the compilation of geothermal maps of Cyprus', I Stylianou et al., *Renewable Energy*, Vol. 88, April 2016, pp.418–429.

[3.13] Airborne Hyperspectral and Thermal Remote Sensing, www.itres.com

[3.14] *Infrared Detectors and Systems*, EL Dereniak and GD Boreman (John Wiley & Sons, Inc., New Jersey, 1996, ISBN 047122092).

[3.15] *Essential Sensing and Telecommunications for Marine Engineering Applications*, Christopher Lavers (Bloomsbury Publishing, London, 2017, ISBN 1472922182).

4

Microwave Sensors

'A possible explanation for the observed excess noise is the one given by Dicke, Peebles, Roll, and Wilkinson (1965) in a companion letter in this issue.'

Arno Penzias' announced detection of the cosmic microwave background radiation, or creation's 'Big Bang' afterglow, in 'A measurement of excess antenna temperature at 4080 Mc/s', *Astrophysical Journal* (1965). Penzias, with co-author Robert Wilson, received the 1978 Nobel Prize for this discovery.

4.1 Problems with visible imagery

The first drawback of visible images is that they only view regions illuminated by the sun. While wintertime images in mid–high latitudes are of limited use, cloud cover is also an issue. Fortunately, the atmosphere is transparent at microwave frequencies, offering good all-weather, continuous daily observations. As discussed in Chapters 1 and 3, little energy is emitted from Earth's surface in the microwave region so instruments are usually active. First we will look at general principles associated with satellite-based microwave systems, introducing Passive and Active systems used for different reasons, and advanced radar methods SLAR, SAR and ISAR for surface imaging. Second, we will examine microwave interactions with water (the hydrosphere), snow and ice (the cryosphere), vegetation, soil, rocks and minerals.

4.2 Passive microwave sensors

Passive microwave radiometry uses the same principle as thermal radiometry in Chapter 3. Most operate between 0.4 and 35GHz. Radiometers measure emitted microwave spectral radiance. Microwave radiance *brightness temperature* $T_B = \varepsilon T$ is given as $T_B = \lambda^4 L_x/2kc$, where symbols given have their usual meaning. Passive microwave sensors are useful, but **spatial resolution is poor** compared with active sensors. The US Defense Meteorological Satellite Program (DMSP) uses passive imaging instruments, such as the Special Sensor Microwave Imager (SSM/I), to monitor snow cover on land, which is hard to map by ground survey because it changes rapidly. Passive microwave radiometry is employed over oceans to determine wind speed to accuracy of 2ms^{-1}, relying on wind speed effects on surface roughness, and thus emissivity. Salinity is determined through its effect on emissivity (figure 4.1).

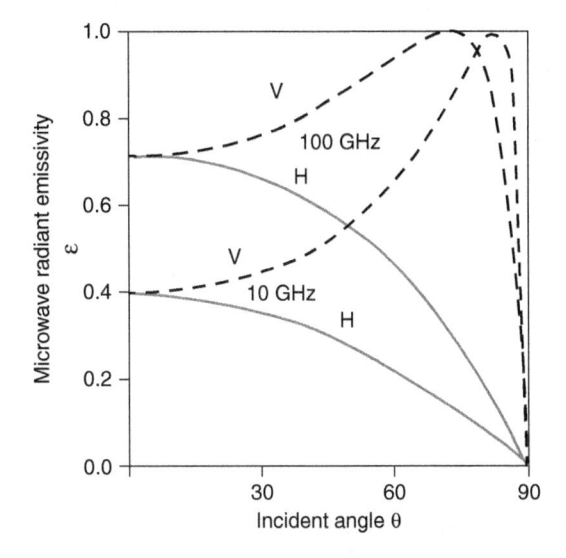

Figure 4.1: *Microwave emissivity as a function of frequency and polarisation.*

AVHRR is another important passive system, a radiation-detection imager for determining cloud cover and surface temperature [4.1]. The term *surface* can mean the Earth's surface or the upper surfaces of clouds. The first AVHRR was a four-channel radiometer on TIROS-N launched in 1978. The most recent instrument version is AVHRR/3, with six channels, launched in 1998 on NOAA-15. Radiometry marks boundaries between ice and water with different emissivities, and measures useful meteorological parameters. If instruments are tuned to atmospheric water vapour absorption, cloud water content, rainfall rate and atmospheric temperature height profiles can be determined.

4.3 Active microwave sensors

We introduce basic radar concepts, starting with the echo principle. Most people are familiar with radar in the context of aviation and maritime safety, the primary sensor used by bridge officers to ensure ship safety. Due to the speed of electromagnetic waves, detection and ranging happen virtually simultaneously. Space-based optical systems cannot provide all the information we require because cloud cover limits use – in fact, worldwide cloud cover varies by 40–80 per cent of surface cover. Hence, sensors that 'see' through weather, available day and night, provide navigational safety advantage. Marine radars are typically X-band (3cm) or S-band (10cm), providing bearing and distance of nearby vessels. Range limitations do not necessarily apply to space-based systems. Radar are not exclusively used; commercial ships integrate a full suite of marine instruments. Radar history is not discussed here, but covered elsewhere [4.2–4.3].

4.4 The echo ranging principle

Most radar, but not all, use the *echo principle*. If the time (t) from transmitted pulse to received echo is measured, knowledge of wave speed helps calculate range.

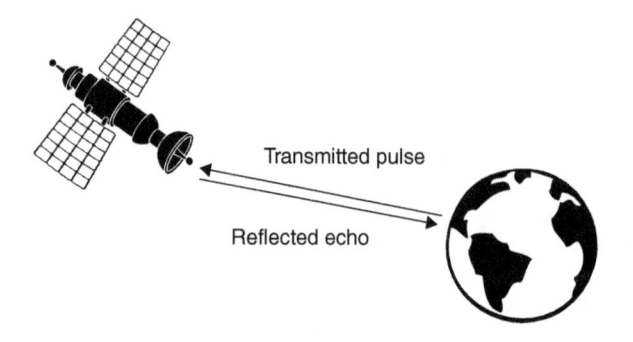

Figure 4.2: *Echo ranging principle.*

Since radar travels at light speed, the distance out and back to a satellite-based radar (figure 4.2) is ct, where t is elapsed time, hence the contact range R will be:

$$R = \frac{ct}{2} \quad \textbf{(eq 4.1)}$$

c doesn't vary much in the atmosphere, and time is proportional to range. This method is known as *pulse delay ranging*. Detailed radar is presented elsewhere [4.2–4.3], but key radar parameters include the number of pulses per second and the pulse duration (τ), typically 0.08μs for high-definition navigation or 3μs for long-range search.

4.5 Radar parameters

4.5.1 Angular resolution

Radar angular resolution is the separation of contacts at similar range on slightly different bearings. Radar can resolve two contacts *if* they have bearing separation more than the horizontal beam width, so radar must not receive echoes from both contacts simultaneously (figure 4.3).

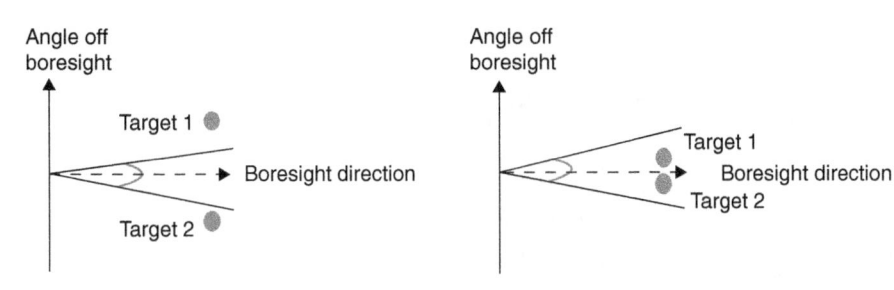

Figure 4.3: *Angular resolution.*

Horizontal a_H and vertical a_V beam widths are given respectively by:

$$a_H = \frac{60\lambda}{D_H} \quad \textbf{(eq 4.2)}$$

and

$$\alpha_V = \frac{60\lambda}{D_V} \quad \textbf{(eq 4.3)}$$

Example 4.1: If a 10cm radar with antenna width = 2m, what is the horizontal beam width?

$a_H = \frac{60\lambda}{D_H}$ Substituting: $a_H = \frac{60 \times 0.1}{2} = 3°$

4.5.2 Range resolution

A pulse leaves a transmitter (figure 4.4). Waves occupy a finite length. The distance occupied by each pulse = velocity × pulse duration = cτ metres. A pulse strikes contact A and reflects, but may continue on to strike contact B.

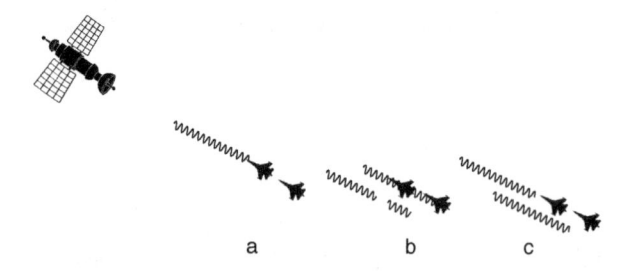

Figure 4.4: *Range resolution (a) outgoing pulse, (b) echoes from targets on the same bearing close together and (c) overlapping echoes from both.*

For two contacts to be resolved, they must produce two echoes – that is, pulses mustn't overlap or they are interpreted as one. Transition from overlap to no overlap is when the separation or range resolution is given by the equation:

$$R = \frac{c\tau}{2}$$ (eq 4.4)

Example 4.2: A navigation radar emits 0.1μs pulses. Find the range resolution and minimum detection range (2 significant figures).

Using $R = \frac{c\tau}{2} = \frac{3 \times 10^8 \times 0.1 \times 10^{-6}}{2} = 15m$. Radar cannot resolve range less than the range resolution, the smallest measurable step. As a radar receiver is switched off while transmitting, range resolution is the radar's **minimum range** as it doesn't process echoes over this time. Range resolution examples for different pulse durations are shown in table 4.1.

Pulse duration (τ)μs	Range resolution (cτ/ 2) m
100	15000
10	1500
1	150

Table 4.1: *Pulse duration vs range resolution.*

4.6 Doppler radar

Some radar, especially SAR, use the Doppler effect.

4.6.1 The Doppler effect

When there is relative motion between radar and target, received frequency differs from that emitted. If range *closes*, received frequency is *higher* than transmitted, and if range *opens*, received frequency is *lower*. The difference between frequencies, first investigated by the Austrian physicist Christian Doppler, is known as the Doppler shift. In 1842, Doppler observed that wave frequency depended on the *relative speed* between source and observer. The Doppler shift is given by:

$$\text{Doppler shift} \Delta f = f_{Received} - f_{Transmitted} = \frac{fV_{REL}}{c} \quad \textbf{(eq 4.5)}$$

In target detection, relative motion between transmitter and target along the direct line between them yields a shift at the target and another shift on echo reception. For echoes, Doppler shift is given by:

$$\text{Doppler shift}_{ECHOES} \Delta f = \frac{2fV_{REL}}{c} \quad \textbf{(eq 4.6)}$$

$f_{Transmitted}$ is the transmitted frequency, giving information about the relative velocity on the direct line between them.

Example 4.3: If the relative velocity between a ship and a vessel is 10ms⁻¹, for 10GHz radar what is the echo Doppler shift (1 decimal place)?

Using: $\text{Doppler shift}_{ECHOES} \Delta f = \frac{2fV_{REL}}{c}$

$$\Delta f = \frac{2 \times 10 \times 10^9}{3 \times 10^8} = 66.7 Hz$$

4.7 Radar antennas

4.7.1 Dish antennas

A common radar antenna is the dish or *parabolic reflector*, whose focusing is similar to a torch's focusing reflector (figure 4.5). If a source is placed at the focal point of a parabolic dish, it produces a directional beam. A sub-reflector usually sits in front of the focus, supported on metal struts. A lightweight sub-reflector is a hyperbola shape with the resulting Cassegrain antenna, figure 4.5a. Cassegrain and Gregorian antennas (figure 4.5b) are used regularly in satellite communications.

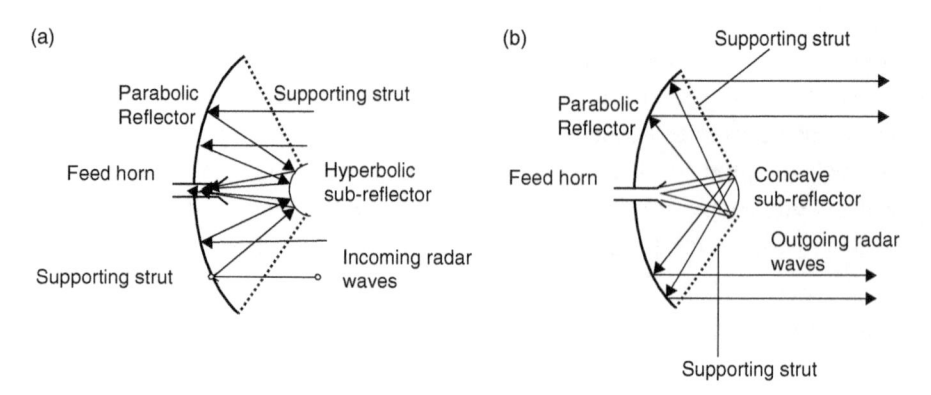

Figure 4.5: (a) *Cassegrain antenna;* (b) *Gregorian antenna.*

Cassegrain antennas use hyperbolic sub-reflectors (convex) while Gregorian antennas have concave reflectors. A parabolic antenna gives conical or pencil beams, narrow vertically and horizontally. Beam width depends on dish diameter given by equations (4.2–4.3). At high frequencies, antenna provide narrow beam width. However, below 5GHz, diameter is too big for mobile use and mobile antennas are restricted to the microwave region.

> **Example 4.4**: A radar dish produces a narrow beam at 20GHz with 3m dish diameter. What is the beam width (1 decimal place)?
>
> $$\alpha_H = \frac{60c}{fD_H} \text{ Substituting: } \alpha_H = \frac{60 \times 3 \times 10^8}{20 \times 10^9 \times 3} = 0.3°$$
>
> Full coverage is obtained by rotating an antenna, although phased arrays offer a better alternative solution, especially in space-based applications.

4.8 Phased arrays

These radar are growing in space-based use and are considered a single fixed aperture but are actually a large array of equally spaced sources covering the area of a single aperture to provide a focused beam. They are steered electronically and don't require mechanical steering, although some radar may do both. Arrays are composed of sub-arrays, each hundreds of phase-shifting elements (passive array) or transmitting emitters (active array). Arrays may be static or rotated. Static radar have no moving parts and use multiple antennas for 360° coverage mounted on ship superstructure. They are lower than mast-mounted radar, restricting search range due to weight. Rotating radar have moving parts and energy is lost at rotating joints (passive only).

4.8.1 Active arrays

Each antenna hole has a small solid-state power supply behind it. Total radar antenna power output is the sum of individual emitter powers. Individual emitter failure results in *graceful* performance degradation with losses much less than passive antennas. A high percentage of active transmitter failure is tolerated due to sophisticated digital software management controlling transmission, but is more complex and costly. To acquire targets, beams are steered with a two-dimensional array (figure 4.6).

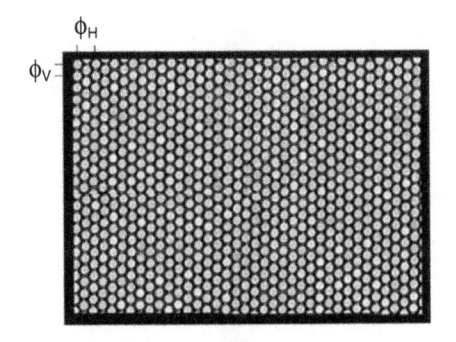

Figure 4.6: *Phase array with regular vertical and horizontal phase shifts between elements.*

Such Active Electronically Scanned Arrays (AESA) use thousands of active transmitters for SAR applications. Beams steer rapidly, allowing ships, satellites or aircraft to use one radar for multiple applications simultaneously, such as navigation, surface detection and aircraft tracking. AESA radar are **multifunctional**, performing many tasks at once. Civilian phased array technology lags behind military use, although it is available in meteorological applications [4.4].

4.8.2 Principle of operation

Consider two sources with a separation *much greater* than the wavelength, so an interference pattern results (figure 4.7).

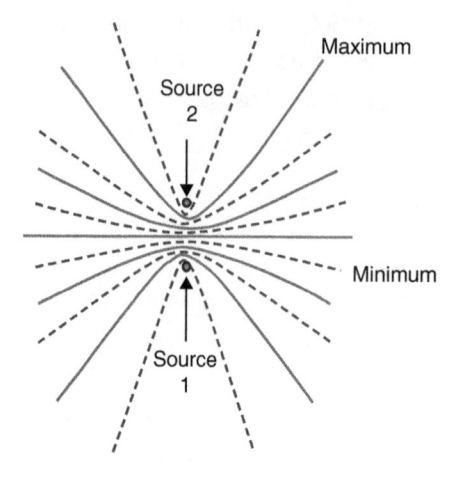

Figure 4.7: *Two-source interference.*

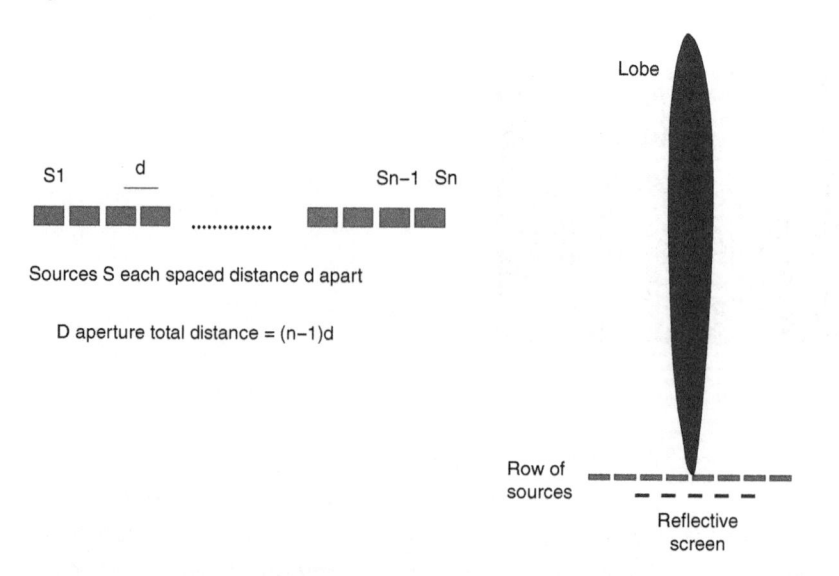

Sources S each spaced distance d apart

D aperture total distance = (n−1)d

Figure 4.8: *Single beam interference.*

If d, the source separation, is *less* than a wavelength, with a suitable microwave reflector behind the sources (figure 4.8), a single beam results.

Horizontal beam width: $a_H = \dfrac{60\lambda}{D_H}$, for half wavelength spacing, ie $d = D = \lambda/2$.

So $a_H = \dfrac{60\lambda}{\lambda/2} = 120°$ is much too large! However, to produce a narrow beam with antenna length D increased, with λ constant and source separation increased above λ, the multiple beam pattern returns. Thus D must increase, but d remains *less* than λ. This contradiction is achieved with a linear array. As more sources are used, overall D increases but d spacing is fixed, giving smaller a_H values. Sources are *in-phase* or have fixed phase between them. In figure 4.8 there is one less space than sources, so $D = (n-1)d$, where n is the source number, and:

$a_H = 60\lambda/[(n-1)d]$ **(eq 4.7)**

Example 4.5: A square array has 2,500 sources. With sources 0.5λ apart, what is the horizontal beam width (2 decimal places)?

Total number of sources = n^2, n is the number of sources in a row or column. As $n^2 = 2,500$ so n = 50

Using: $a_H = 60\lambda/[(n-1)d]$ and substituting:

$a_H = 60\lambda/[(50-1)0.5\lambda] = 60/[(49)0.5] = 120/[(49)] = 2.4°$

4.8.3 Linear array beam steering

If a time delay is introduced across each source (right to left), they radiate slightly later than previous ones. Thus the wavefront tilts because energy travels further from the first, and progressively less from subsequent sources (figure 4.9, left). If a time delay is introduced in the opposite direction, the beam tilts in the *opposite* direction (figure 4.9, right).

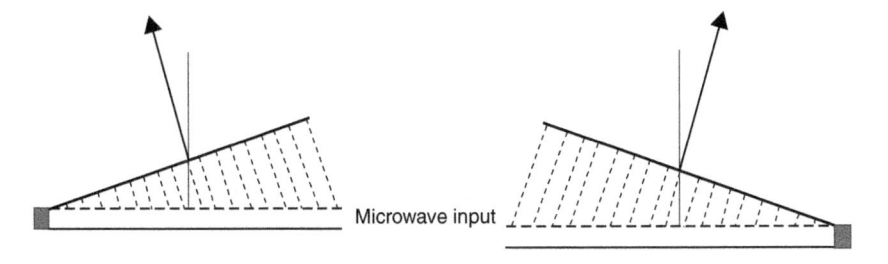

Figure 4.9: *Beam steering off boresight (left), beam reversal (right).*

The beam has a time delay equivalent to its phase change ϕ. The phase difference between elements is given by:

$\phi = 360 d \sin\theta / \lambda$, d is source separation and θ the angle the beam steers from its original position. Changing ϕ produces a beam in a specific direction. d is given as $d = k\lambda$ with k between 0.5 and 0.9, so ϕ becomes:

$\phi = 360 \, k \sin\theta$ **(eq 4.8)**

Example 4.6: A radar operates with $\lambda = 0.1$m and is steered electronically, changing phase of one emitter relative to the next. If the beam steers $\theta = 60°$ off boresight, what angle ϕ is applied between the emitters? k = 0.5 (1 decimal place).

Using equation (4.8):

$\phi = 360 k \sin\theta = 360 \times 0.5 \times \sin 60 = 155.9°$

4.8.4 Phased array radar advantages

Beam steering is fast, typically microseconds. Antennas produce hundreds of beams per second, for different locations. Some beams produce search patterns, others can track vessels or give guidance signals. Arrays perform multiple functions, otherwise requiring separate radars. Due to the many beams produced, data rate is high, yielding long Maximum Detection Range (MDR) and high signal/noise ratio improvements. Antennas are mounted on ship superstructure as there is no need for mechanical antenna movement. Recent arrays can rotate and are used in space-based operations. The main phased array disadvantages are cost and complexity, which limit use on civilian vessels.

4.9 Radar imaging

SLAR is an imaging radar. SAR is a high-resolution SLAR refinement, while ISAR considers target motion with reference to radar to generate larger effective aperture sizes. One way to think of a radar image is a map of moisture content and surface morphology (shape). A rough surface reflects more energy back to sensors than a smooth one, but what do we mean by *rough* or *smooth*? In radar, these terms depend on wavelength: see Rayleigh's criterion (Chapter 2).

Different common wavelength bands used are: 3cm (X-band) 10GHz, 5cm (C-band) 6GHz, and 25cm (L-band) 1.2GHz.

Chosen wavelength depends on the information users want. Radar reveals differences in surface relief when surface variation is **more than** the wavelength. Consider a lake with small waves below 10cm. At a wavelength of 25cm (L-Band), the lake appears smooth. Most energy is reflected *away* from a sensor by the surface (specular reflectance) and little backscatter is detected. However, if the same lake, with the same 10cm waves, is sensed by 5cm radar (C-Band), the lake appears rough. Radar reflects in many directions (diffuse reflectance) and more backscatter is detected. The incident angle and the surface (rough or smooth) affects how radar interacts. Radar maps sea ice, as sea water returns are different from ice-covered seas, and reveals oil spills because oil-covered water has a smoother surface; for example, ship wakes make the sea rougher. Bright images reveal surfaces that reflect well. Ship superstructure can be a good reflector, reflecting strongly back to space-based sensors. Urban tin shacks in Durban, South Africa, are visible from space. Where vegetation or soil is dry, microwaves penetrate using long wavelengths, revealing features hidden in the visible. Images are harder to interpret than photographs, with two problems: **layover** and **speckle** (see 4.9.2). Many active radar satellites are in use, providing vital maritime information, such as Jason-2 for operational oceanography. Ocean circulation is studied by measuring sea level height [4.5–4.6].

4.9.1 Side-Looking Airborne Radar (SLAR)

SLAR is an active **real aperture** technique developed in the 1950s for military reconnaissance from Second World War PPI radar. SLAR looks to one side of the flight direction to produce continuous surface strip-maps. It transmits short radio frequency (RF) pulses. Antennas are long and thin, and mounted with their long axis parallel to the platform motion. Resolution in the direction *parallel* to the platform's direction (the **azimuth** or along-track direction) is achieved by virtue of antenna length L. Good range resolution is proportional to pulse duration and perpendicular to flight direction.

If we write beam width in radians $\alpha \approx \lambda/L$, azimuth resolution is:

$R_a = H\lambda/(L\cos\theta)$ (**eq 4.9**),

where θ is the incidence angle (figure 4.10).

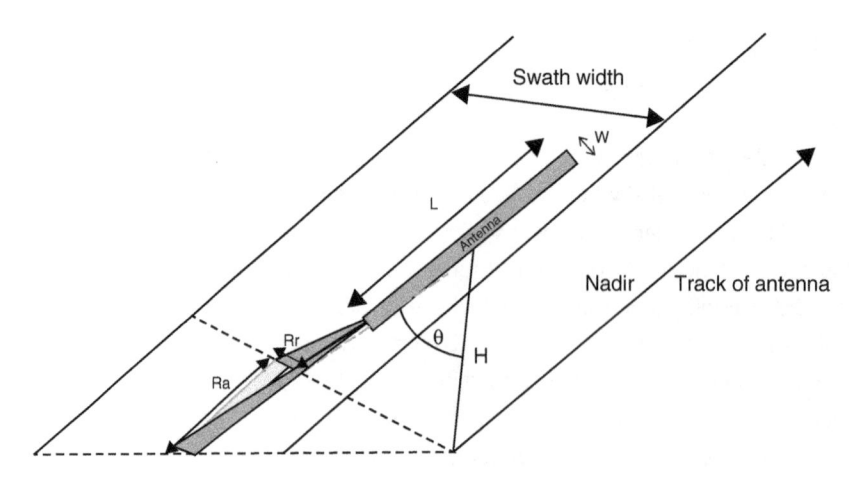

Figure 4.10: *Geometry of a SLAR system.*

The antenna is length L, width w, situated H above a surface. It views to one side. Ra is resolution of azimuth along-track direction at angle θ. Resolution in the perpendicular or *cross-track direction* is determined by pulse duration τ. The condition for two points to be resolved is that their distances must differ by >cτ/2. The view from a plane, or a satellite, is slanted down towards sea and land surfaces (see figure 4.4). Slant range resolution is Δs, so ground range resolution:

Rr = cτ/(2sinθ)   **(eq 4.10)**

Azimuth resolution Ra is proportional to H. Although SLAR can achieve good resolution from airborne platforms, it is inadequate for satellites.

Example 4.7: Consider a SLAR system, λ = 2cm with antenna 6m long and 30ns pulse duration from an aircraft at 4km. Using equations (4.9–4.10), 6km from the ground track θ = 56.3°, Ra = 24m and Rr = 5.4m. 20km from the ground track θ = 78.7°, so Ra = 68m and Rr = 4.6m. With the same SLAR system at 200km altitude, 6km from the ground track θ = 1.7°, Ra = 66.7m and Rr = 151.7m. 20km from the ground track θ = 5.7°, so Ra = 67m and Rr = 45.3m.

4.9.2 Image distortions

SLAR measures range to scattering objects within its instantaneous footprint. In the simplest processing, images are presented so **slant range** increases uniformly across imagery. This is a form of distortion, since we actually require ground range to increase uniformly, but can be corrected.

Two further problems are layover and shadowing:

Layover: This is where the top of a vertical object is closer to a radar than the bottom, and top regions can overlay the returns from the base, such as city towers. Layover is significant at small incident angle θ.

Shadowing: This is a big problem at large θ values. Part of a surface can be hidden from radar by another surface in front. No signal returns from the shadow regions, and the corresponding image is dark. Figure 4.11 illustrates these distortions.

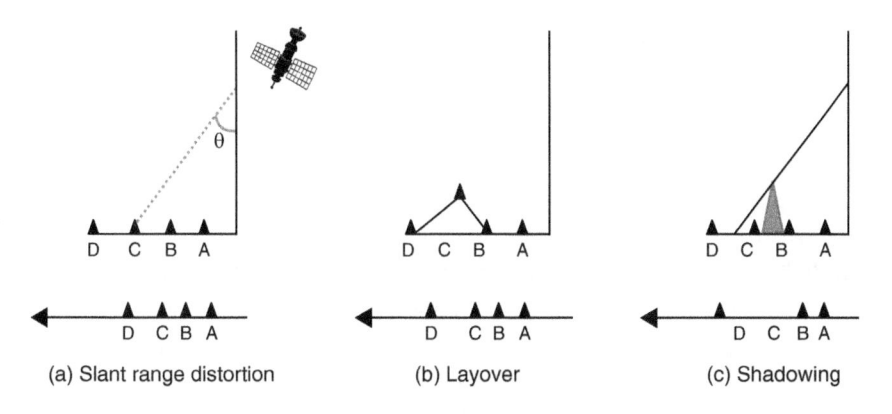

(a) Slant range distortion (b) Layover (c) Shadowing

Figure 4.11: *Various surface representation distortions.*

Image speckle is reduced by averaging returns, as image intensity statistical variance is reduced greatly in inverse proportion to the number of averages.

4.9.3 Synthetic aperture radar

Maritime radar was developed to detect ships, but there was also a need to develop higher resolution systems and techniques for radar platforms at a great distance from targets, especially satellites. SAR uses platform movement to make antenna length appear longer than it is. Signal processing techniques produce the effects of a long antenna and thus narrow beam width, without increasing real length (figure 4.12), hence the term 'synthetic aperture'. This apparent increase in size is achieved using *relative motion* between radar and target, using the **Doppler shifted signal**. Effective antenna beam width can be reduced, so radar along-track resolution is comparable with range resolution. SAR techniques produce beam widths of under 0.1°, generating high-resolution surface maps. It now aids ship and oil rig platform identification from space. So how do we get better angular resolution? The answer is to use a larger antenna aperture *D* for the antenna. However, increasing antenna size until we get better resolution is impractical. Instead, we *simulate* aperture length by moving a fixed antenna along a path relative to the object.

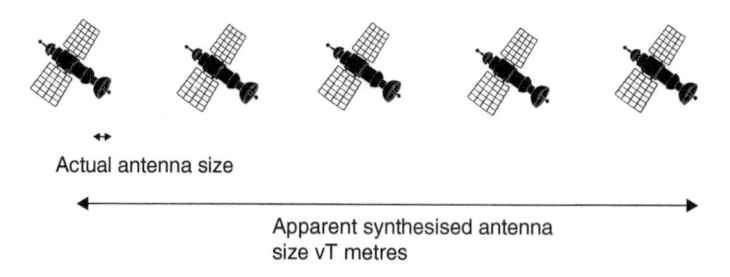

Actual antenna size

Apparent synthesised antenna size vT metres

Figure 4.12: *SAR integration with satellite in image.*

SAR takes advantage of radar's long range, low attenuation properties. It is less affected by atmospheric and weather conditions compared with other systems. SAR works like phased array but instead of having many parallel Transmitter/Receiver (T/R) antenna, it uses a single T/R antenna, which transmits and receives over a common path. With phased array, beams form by switching emitters on at different times. A beam pattern is the result of a platform moving across a target with velocity V and storing echoes as amplitudes and phases over a time period *T*. A digital SAR processor reconstructs signals otherwise obtained from an antenna with aperture length so higher resolution is achieved (figure 4.13).

SAR = VT   (**eq 4.11**)

Example 4.8: A surveillance satellite operates AESA radar in SAR mode. The satellite travels at velocity V = 7.7kms^{-1} and transmits at 3GHz. The satellite platform overflies North Korea and processes returned signals over a time period of T = 2s. What is the theoretical synthesised aperture D length in metres (3 significant figures)?

SAR synthesised aperture length = V × T = 7700 × 2 = 15.4km

SAR echoes from each pulse are recorded by a processor. As the platform moves over a ship, *all* ship echoes for each pulse are recorded for the entire time the ship is in the beam. Like a phased array, it is important SAR signals are *coherent* – that is, they must be at a single frequency with constant phase difference, so phase shouldn't fluctuate while the SAR scans over the target.

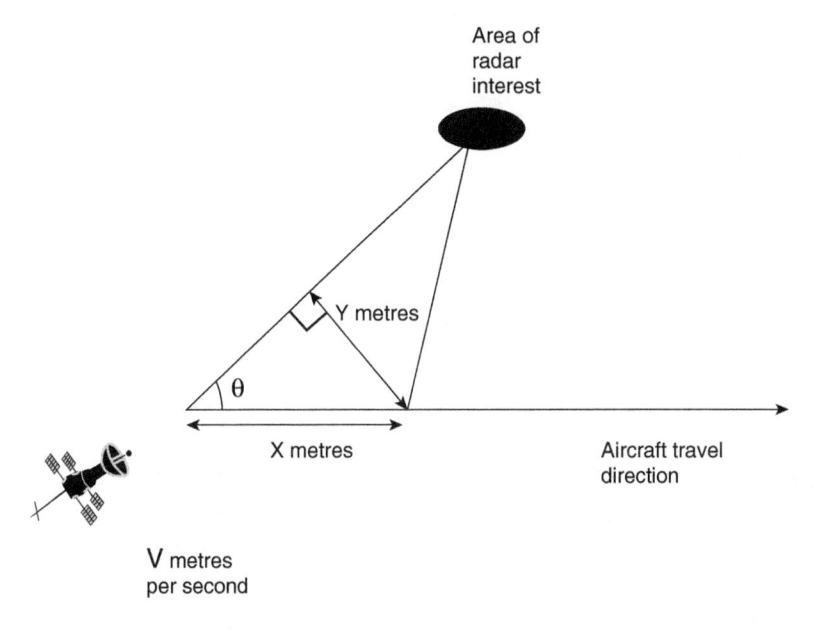

Figure 4.13: *Synthetic aperture of radar over integration period.*

For 3GHz (10cm), a real 0.5m antenna gives a 3.6° beam width. With the satellite SAR acquiring a data track on a ship for 2s at 7.7kms^{-1}, we fly in the direct line distance X, here 15.4km (figure 4.13). Echoes received during this time are integrated, giving an apparent antenna larger than the real thing. Allowing for observation angle, if the angle was 30° this gives Y = 7.7km effective aperture. A 7.7km antenna gives a beam width of about 7.8 × 10^{-5} degrees, an attractive technique for satellites orbiting Earth at thousands of miles per hour.

The first civilian space-based SAR (SeaSat), launched in 1978, was the first to produce data demonstrating the value of satellite SAR. SeaSat details are found elsewhere [4.7]. Satellite SAR will form a key part in future maritime surveillance and ocean satellite missions. ESA's polar mission, Envisat, and Earth Observing System (EOS) included SAR among their payloads. SAR is similar to Doppler processing – a given along-track co-ordinate on the sea surface has a unique time variation of frequency associated with it. As long as signal amplitude and phase are recorded, Doppler components are obtained. The process is repeated for other values of along-track co-ordinates. With SAR, consider a strip vertically below at height

H. If the real SAR antenna length is L, beam width in radians is around β~λ/L, so arc length is λH/L along the surface track. Looked at differently, a given point on the ground is illuminated by an antenna *only* while the SAR radar travels this arc distance, the maximum synthetic aperture length. Radar synthetic aperture angular resolution βsynth = λ/(max synthetic aperture length) = λ/(λH/L)≈L/ H, so surface resolution = βsynth × H = (L/H) × H, giving surface resolution of ≈L, and independent of H, unlike SLAR.

4.9.4 SAR imaging targets

If targets move, the complex way that data is processed results in an apparent object position shift. As radar approaches a stationary target, the Doppler shift decreases, reaching zero when radar has the same along-track position as the target. There is now no relative velocity component directly between source and target. If the target moves, a *second* shift adds to the platform motion. This means the Doppler shift is zero at a *different* along-track co-ordinate (different relative velocities). Imaging radar is widely applied in oceanography. Surface wave fields are imaged distinctly. Wave diffraction from coastal features and refraction from bottom topography is visible, and used to look at sea ice. Boundary delineation between ice floes and open water is easy. Across-track resolution is achieved with techniques such as pulse compression, with 500MHz bandwidths, leading to 0.5m resolution. Two SAR techniques used are *swath mapping* – ultimate resolution is limited to half real antenna length, and *spotlight* – a radar beam is trained on a patch of ground for perhaps tens of seconds, providing centimetric resolution.

4.9.5 SAR maritime monitoring

SAR has the advantage of being 24/7 (active rather than passive solar electro-optical) and is able to penetrate thick cloud and fog, providing high-resolution surface imagery. SAR modes include vessel and oil spill detection and sea ice and iceberg detection, discussed in Chapter 6. Surveillance is discussed elsewhere [4.3], but maritime use is introduced. Surveillance radars are vital to modern naval fleets being able to detect small targets in rough seas. The SeaSpray 5000E radar (Leonardo) is a compact AESA with air to surface cover, on fixed and rotary wing platforms, combining mechanical and electronic scanning for long-range small target search [4.8]. A real-time SAR processor provides target and vessel imaging and a range-Doppler map. Results are outstanding, with near-photographic quality resolution of surface water and land terrain, identifying features to 200nm.

Doppler components are greater from upper masts, but most returns are from low bulkhead and hull components moving with small Doppler components. The SeaSpray 7500E X-band system SAR imagery can detect 320nm range, providing capabilities as yet only dreamed of in the civilian merchant and aviation arena. In 2017, Selex ES supplied X-band 200nm range Osprey multimode surveillance radar to Norwegian Search and Rescue helicopters for land surveillance, strip and spot SAR ground mapping, MTI and SART beacon detection, imaging and classification.

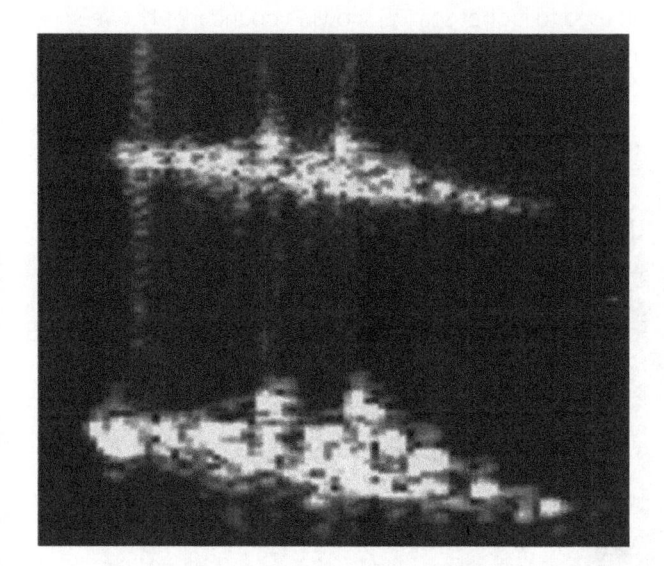

Figure 4.14: *ISAR image of two ships. Courtesy Leonardo Airborne and Space Systems Division.*

The first Shuttle Imaging Radar (SIR) experiment was carried out on board the shuttle in 1981 (SIR-A). Over the last 30 plus years, several systems have been launched with increasingly sophisticated SAR instruments. In discussion with Leonardo, it isn't a straightforward matter to take a maritime SAR system and customise it for space-based applications! The first European SAR ERS-1 was launched in 1991, TOPEX/Poseidon in 1992, besides RADARSAT, and the Almaz series also carry radar. Data from satellite radar investigates oceans and ice sheets, and monitors global climatic change.

At first look, SAR is indistinguishable from SLAR, but it overcomes problems set by SLAR altitude-dependence on H (equation 4.9). The same geometry applies for both in that they emit pulses and analyse echoes to obtain resolution in the range direction. Higher resolution in azimuth along-track is achieved by sophisticated echo processing. Unlike SLAR, SAR relies on platform motion to achieve high resolution in the azimuth direction.

4.9.6 SLAR and SAR images

SLAR and SAR images appear like aerial black-and-white photos, singly or as stereo pairs. Imaging radar is used in oceanography to image surface wave fields, as is wave diffraction by coastal features and refraction from bottom topography. Imaging radar has also been used to look at sea ice, showing boundaries between ice floes and open water and allowing comparison of consecutive images to track ice floes.

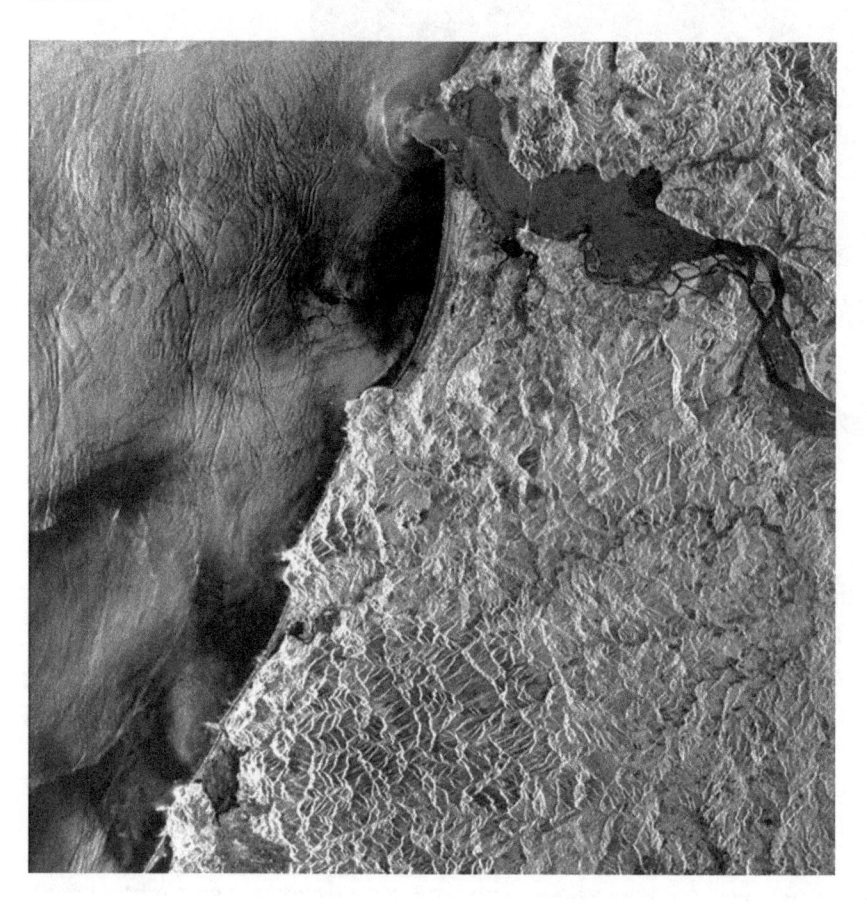

Figure 4.15: *Washington-Oregon coast: digitally processed 2013. Acquisition date: 10 August 1978. This image shows influence on wave patterns of a subsurface sediment spit near the mouth of the Columbia River between Oregon and Washington. Strong returns appear bright. Processing organization: ASF. Author NASA/JPL.*

4.9.7 Polarisation and 3D images

Most conventional radars operate with a single polarisation, horizontal in horizontal out (HH), or vertical in vertical out (VV), so only part of the full electromagnetic scattering response is measured and some target information is ignored. There are also cross-polarisation products – that is, horizontal in vertical out (HV) and vertical in horizontal out (VH). Complex targets are made of scatterers that are polarimetrically sensitive, giving them unique signatures that allow more reliable classification. 3D images are generated using two radar antennas mounted one above the other in an interferometric arrangement. For flat ground, phase differences between antennas are a known function. If an object is located on flat ground, this relationship changes in a way that depends on object height, so 3D radar ground representation may be constructed.

4.10 Inverse Synthetic Aperture Radar (ISAR)

ISAR is a type of SAR to image targets like ships, aircraft and space objects. ISAR imagery is generated from data collected as **targets rotate** in the radar beam. If a target is uniformly illuminated, and processing performed on data collected while the target rotates through a small integration angle ψ, the slant range and Doppler frequency of scatterers at the target angle extremes shift *less* than slant range and Doppler resolution processed images consist of estimates of sizes and scatterers' positions in slant and cross-angle. ISAR permits remote surveillance of land-based movements, remote ship target classification, and differentiation between military and civil targets. Target animations over several seconds can correctly classify targets on the basis of appearance and behaviour.

4.11 Sensor system types

Real missions include components of passive and active radar and other systems, such as the ERS-1–2 satellites, Tropical Rainfall Measuring Mission (TRMM) and Soil Moisture Ocean Salinity (SMOS) satellite.

4.11.1 The European Remote Sensing (ERS) Satellite

ERS-1 was launched by ESA in 1991 into a 770km near polar orbit, collecting information on areas difficult to observe, such as oceans and ice-covered areas, producing surface images in all weathers, 24/7/365. Microwave sensors on ERS-1 included an Active Microwave Instrument (AMI) and Radar Altimeter (RA).

AMI produced radar images, operating in three ways: *SAR image* mode, *wave* mode to measure ocean parameters, and *wind scatterometer* mode for speed and direction at sea surfaces.

RA An altimeter determines surface height, vital as the satellite's position must be known. Corrections are made as pulses are affected by transit through the atmosphere. Interactions between microwaves and surfaces are complex, making it difficult to use over land. It provides accurate Sea Surface Height (SSH) measurement to centimetres. Wave height is important for shipping safety and oil-rig operation. Winds generate waves, varying SSH. Altimeter data determines wind speeds over oceans, important for forecasting. Topography or sea surface shape mirrors the underlying sea floor, creating 'hills' over oceanic mountains as gravity is stronger above them (figure 4.16). The ERS satellites studied changes in ocean height over currents and detected small eddies. One disadvantage is that the altimetry ground tracks only cover several narrow bands of Earth's surface, resulting in coverage gaps. Longer repeat cycles give greater ground coverage, providing information about more of Earth's surface in each cycle, essential to record differences in the level of the sea over wide ocean areas. Another disadvantage is short-term events such as depressions, which influence SSH, and change too rapidly to be studied. The ERS altimeter looked at sea-surface topography anomalies in the 1997 El Niño event, over months. Space-based RA can measure the sea ice freeboard height and other surface factors (figure 4.16).

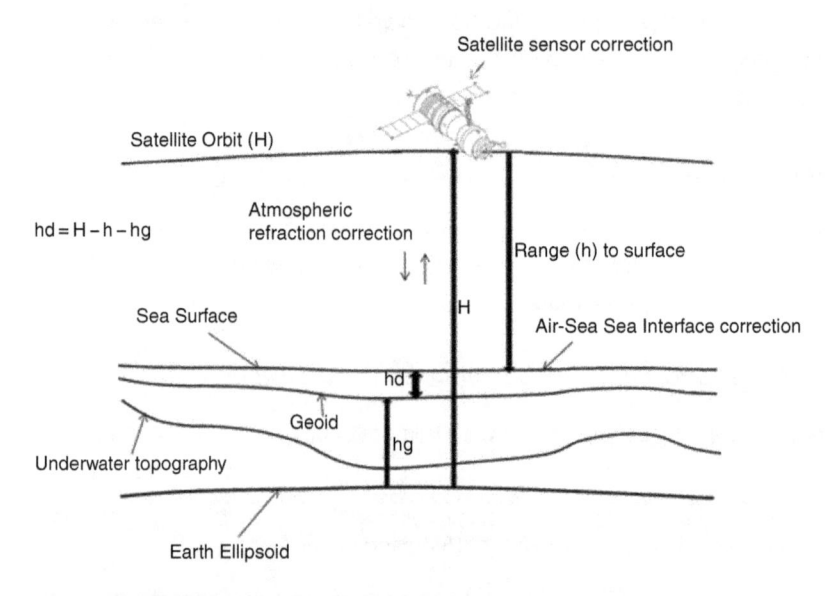

Figure 4.16: *Radar altimetry and measurement of geoid values.*

ERS–1–2 provided tandem interferometric over time as black-and-white interference intensity images.

4.12 Interaction of microwaves with different surfaces

Echoes from different surface features in processed images depend on the energy returned from individual scattering targets. Backscattered intensity depends on how radar interacts with the surface, which depends on several factors. These include contributions of the radar system itself (frequency, polarisation and viewing geometry) and surface characteristics (land cover type, moisture content, topography and relief). Many of these are interrelated, which makes it hard to separate contributions. Changes in these factors impact the response of others. We can group these characteristics into three areas, which control radar energy/surface interactions, namely: surface roughness, radar viewing angle and surface geometry, and moisture content and electrical properties (permittivity).

4.12.1 Interactions of microwaves with water

Surface conductivity has a large effect on radar backscatter as a surface like metal or salt water has a greater conductivity than a non-conducting surface such as rock. Conductivity depends on surface electrical properties, the complex dielectric constant $\varepsilon = \varepsilon_r + i\varepsilon_i$. The real part of ε_r ranges 60–80. The water content in soil or vegetation affects returns *much more* than vegetation type or soil texture. Water is associated with low returns, while land and vegetation cover are associated with high returns. Radar scatter is sensitive to dielectric constant changes and is useful for mapping soil moisture, snow pack (wet or dry) and wetlands, where cloud cover, often occurring in coastal regions, doesn't affect SAR imagery.

Satellite inclination towards a flat water surface has a major effect on returns. If the sensor is forward looking, reflected energy is low as scattering is negligible and waves reflect in the forward direction. If a surface is rough, more energy backscatters towards the satellite. If the surface is very rough, significant cross-polarisation (HV and VH) occurs and appears bright on imagery. If radar points directly down (*nadir*), an altimeter records a moderate signal directly reflected even though water has strong absorption. Clearly, operation at water molecule resonances such as 2.45GHz must be avoided! In the microwave radiometry sense, water and metals emit least (*cold*). Rocks and sandy surfaces by contrast are 'hot', with emissivity typically 0.9–0.95. Care must be taken with targets over sea surfaces, as near-surface winds at the surface make it rough and increase backscatter.

4.12.2 Interactions of microwaves with vegetation

Vertical polarised microwaves are sensitive to vertical structures on the surface, such as tall buildings and, on a smaller scale, trees and crops. VV returns provide good contrast among vegetation types and canopy structures. Cross-polarised returns (HV or VH), representing depolarisation, result from multiple reflections within vegetation. HV and VH images are more sensitive to crop structure within total canopy volume, and provide information that is *complementary* to HH and VV images. Polarisation helps derive crop condition information, and estimates crop yields early in the growing season. Multi-temporal images map changes during the growing season and land use changes. The C-band is sensitive to scattering elements and measures the upper part of tree canopies. RADARSAT-2 was used for forest mapping and detection of structural differences between forests, and cross-polarisation images monitored burned forest distinctly from unburned. Whereas optical satellite images give observers images of surface distribution of leaf canopy chlorophyll components (NDVI and NDVI-derived products), microwave images provide information more about the temperature regime and spatial distribution of electrolytes inside the overall arrangement of bulk leaves.

4.12.3 Interactions of microwaves with soil

Horizontally polarised (HH) penetrates canopies more than vertically polarised. HH returns provide more information about soil conditions beneath vegetation. Soil moisture is an important factor for natural resource applications such as hydrological modelling and agricultural practices. Extraction of soil-moisture measurements is possible with microwave measurements. The physical basis for soil moisture determination is that, as moisture *increases*, soil dielectric permittivity *increases*, while emissivity and brightness temperature decrease. Reflectivity recorded by active microwave instruments is unaffected by emitted signals, so active and passive instruments are used for different aspects. Complicated models have been developed that account for temperature, moisture profiles, soil types, and moist and rough vegetation. Humus, a key component of tilled soil, affects soil structure and physical properties. Small amounts of humus will affect soil hydro-physical and mechanical properties. Soil moisture data is also retrieved from active space-borne microwave observations [4.9].

4.12.4 Interactions of microwaves with rocks and minerals

A combination of measurement techniques has investigated dielectric properties of 80 rock samples in the microwave, measured in 0.1GHz steps across 0.5–18 GHz

[4.10]. Rocks have high emissivities, typically 0.9–0.95. RADARSAT-1 imagery provides geologists with useful information for mapping structure and geomorphology, as well as stereo visualisation techniques for geo-hazard characterisation. Researchers have shown that the dielectric loss factor decreases with increasing frequency for most rocks [4.10]. Other workers have shown that mineral dielectric properties can be identified [4.11]. Multi-temporal images map changes in landslide distribution and facilitate geomorphology and slope mapping. RADARSAT-1 provided terrain information such as surface roughness, for understanding processes such as bedrock weathering. In cross-polarisation, backscatter from bedrock fracture zones and fault scarps is highlighted and contrasts against surrounding unpolarised regions. Integration of RADARSAT-1 and Landsat TM imagery for mineral exploration in Egypt has been explored [4.12], and over arid sedimentary areas [4.13].

4.12.5 Interactions of microwaves with snow and ice (the cryosphere)

4.12.5.1 Sea and land applications

Remote sensing permits mapping of land ice, snow, and global sea ice. It is achieved with several systems, including passive microwave sensors, active radar altimeters and SAR. Microwave systems have the advantage of data collection through clouds and in the dark. In high latitudes, winter months are dark with few daylight hours and temperatures are so low that seas freeze. The Antarctic land mass is covered with the world's largest ice sheet, and around the continent's edge seas are frozen all year but the *amount* of sea ice increases in winter, extending into the Southern Ocean.

Snow depth, grain size and pack structure are related to backscatter in a complex way. SAR backscatter is influenced by several factors including roughness, topography, land cover, incident angle and moisture content. Use of polarimetric SAR imagery provides more accurate information on snow packs. Through the operation of RADARSAT satellites, Canada has become a world leader in operational sea ice monitoring. Identifying boundaries between ice and open water is the primary task for sea ice mapping. Backscatter contrast between ice and water determines the ice edge definition, vital for shipping. Icebergs, like ships, also show up in C-band radar images as bright point targets. Cross-polarisation data helps iceberg detection, especially at steep incident angles, often encountered with space-based SAR. Full polarimetric radar systems can improve discrimination between icebergs and ships. Common snow parameters are estimated by SAR. Current and planned all-weather high-resolution SAR monitoring systems include

ALOS, HJ-1C, RADARSAT, RCM, RISAT 1–2, Sentinel-1, CSG, CSK, KOMPSAT-5, Meteor M and MP, PAZ, Terra-SAR-X and TanDEM-X.

Questions

4.1 State reasons for growth of microwave alongside optical satellite sensors for maritime and littoral applications, and discuss the differences between existing active and passive systems.

4.2 Explain how SAR works, and achieves significantly smaller beam width than SLAR, and why this is important for satellite-based platforms.

4.3 Discuss the importance of radar imaging of moving targets and how this can be achieved. Explain the differences between across-track (slant) and along-track, the azimuth direction of the flight itself or azimuth resolution. Consider first a SLAR system operating at $\lambda = 3$cm with an antenna 2m long and 60ns pulse duration, from an aircraft at 5km height. Find Ra and Rr for: (a) 5km and (b) 25km from the ground track. If the same SLAR system were operated at a height of 220km, find Ra and Rr for: (c) 5km and (d) 25km from the ground track. Comment on your results.

4.4 A radar operates at 9.54GHz with 2.5m dish size. What is the horizontal beam width?

4.5 A search radar emits 2.5µs pulses. What is the range resolution and minimum detection range (2 significant figures)?

4.6 A square array has 900 regularly spaced sources, 0.5λ apart. What is the horizontal radar beam width (2 decimal places)?

4.7 A radar surveillance satellite operates in SAR mode, travelling at 7.9kms^{-1} and transmits on 2.9GHz. The satellite platform overflies Iran and processes returned signals over a period T = 2.5s. What is the theoretical synthesised aperture D in metres (3 significant figures)?

4.8 If the relative velocity between a ship and another moving platform is 288ms^{-1} at 3.1GHz radar, what is the resulting echo frequency (1 decimal place)?

4.9 A radar operates with $\lambda = 0.1$m and is steered electronically, changing phase angle of one emitter relative to the next. If the beam is steered 60° off boresight, what phase angle φ must be applied between adjacent emitters, k = 0.5 (1 decimal place)?

4.10 Explain the origins of the passive *microwave* emissivity measurement method in terms of *thermal* radiometry methods.

REFERENCES

[4.1] noaasis.noaa.gov/NOAASIS/ml/avhrr.html

[4.2] *Essential Sensing and Telecommunications for Marine Engineering Applications*, Christopher Lavers (Bloomsbury Publishing, London, 2017, ISBN 1472922182).

[4.3] Reeds Marine Engineering and Technology, Vol 14, *Stealth Warship Technology*, Christopher Lavers (Thomas Reed, London, 2012, ISBN 9781408175255).

[4.4] The National Severe Storms Laboratory, www.nssl.noaa.gov/

[4.5] Bibliothèque des projets du CNES, jason.cnes.fr

[4.6] TOPEX/Poseidon project, which obtains similar information, sealevel.jpl.nasa.gov/missions/topex/

[4.7] Alaska Satellite Facility, SeaSat images and history, www.asf.alaska.edu/

[4.8] SeaSpray 5000E, Leonardo, www.radartutorial.eu/19.kartei/08.airborne/karte012.en.html

[4.9] 'Soil Moisture Retrieval from Active Spaceborne Microwave Observations: An Evaluation of Current Techniques', BW Barrett, E Dwyer and P Whelan, *Remote Sensing* 1(3), 2009, pp.210–242.

[4.10] 'Microwave dielectric properties of dry rocks', FT Ulaby et al., *IEEE Transactions on Geoscience and Remote Sensing*, Vol. 28(3) (June 1990), pp.325–336.

[4.11] 'Integration of Radarsat-1 and Landsat TM images for mineral exploration in East Oweinat district, South Western Desert, Egypt', TM Ramadan, AH Nasr and A Mahmood, www.isprs.org/proceedings/XXXVI/part7/PDF/236.pdf

[4.12] 'Microwave signatures over carbonate sedimentary platforms in arid areas: Potential geological applications of passive microwave observations?', C Prigent et al., *Geophysical Research Letters*, Vol. 32 (2005) L23405, 2005, doi: 10.1029/2005GL024691.

[4.13] National Snow and Ice Data Center (NSIDC), nsidc.org/arcticseaicenews/

5
Atmospheric Interactions with Electromagnetic Radiation

'For the first time in my life I saw the horizon as a curved line. It was accentuated by a thin seam of dark blue light – our atmosphere. Obviously this was not the ocean of air I had been told it was so many times in my life. I was terrified by its fragile appearance.' Dr Ulf Dietrich Merbold, the first West German citizen to have flown in space

5.1 Radiation from the sun and the solar radiation spectrum

In Chapters 2–4 we examined interactions of electromagnetic radiation with Earth's common surfaces. In addition to surface interactions, electromagnetic radiation must make one or two transits through Earth's atmosphere, whether from a natural or man-made source. Consequently, the atmosphere and, for maritime Earth observation, the ocean as well, will modulate **in**cident **sol**ar radi**ation** (insolation). As radiation propagates through the atmosphere or sea, it interacts with the media it encounters in transit, through scattering (energy is not lost but merely redirected) or absorbed (energy is lost to the media, heating it up, which is then re-radiated in different directions at longer wavelengths).

Three main gases – water vapour, carbon dioxide and ozone – are efficient atmospheric solar radiation absorbers. Insolation is the main natural radiation source for passive remote sensing attenuated through the atmosphere. Energy is lost from the forward direction and the combined effect of scattering and absorption is termed 'attenuation', expressed as an **extinction coefficient**, which evaluates energy reaching Earth's surface as a ratio of incident radiation upon the atmosphere's outer limits of transmittance (see figure 5.1).

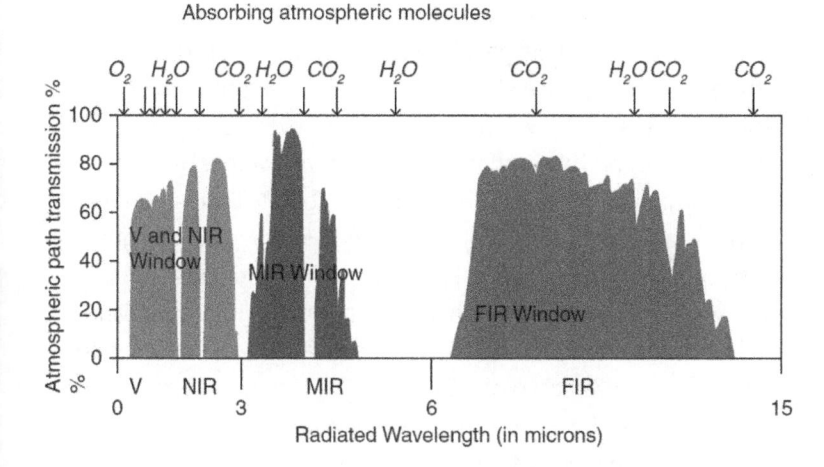

Figure 5.1: *Atmospheric transmission as a function of wavelength.*

The atmosphere has a refractive index a little higher than the vacuum of space. Solar radiation reaches the outer limits of Earth's atmosphere from space undiminished except for geometric spreading effects. The amount passing through a unit area is *inversely* proportional to (distance)^2 between the source and across the area considered. Earth is 93 million miles from the sun, and intercepts about 10^{-8} of total solar output! Terrestrial observation of solar radiation is difficult due to high atmospheric losses at sea level, where most measurements are made. UV radiation rockets have extended measurements to short wavelengths where solar radiation is totally blocked by atmospheric absorption. However, the best place to evaluate solar spectra is outside the atmosphere entirely. Recent satellite monitoring systems, such as SOHO and Cluster II, were launched to observe solar output and monitor Earth's magnetic field respectively, and have greatly advanced solar monitoring. Future planetary missions will use rocket and balloon sensing systems.

5.2 The atmospheric absorption spectrum

As we know, accurate incident radiation measurement is affected by energy lost to space by scattering, with some absorbed. Most radiation penetrating the atmosphere lies between 3 and 50μm, which absorbs direct solar radiation, and long-wave (terrestrial) radiation, as discussed earlier. Absorptivity variation for atmospheric constituents across the spectrum is shown in figure 5.2 (see plate section).

There is strong variability observed between wavelengths, sharp absorption peaks separated by abrupt dips where gas absorption is low. At the short end of the spectrum, the atmosphere effectively absorbs almost all incident radiation. However, it is largely transparent between 0.3 and 0.7μm. Beyond this, there is a sequence of sharply defined absorption bands alternating with relatively transparent regions, which are the most useful 'windows' for satellite measurements, avoiding bands where atmospheric absorption is strong. However, wavelengths of maximum absorption may be chosen for space-based meteorology studies for exactly this reason. For example, atmospheric depth soundings by multiband radiometers exploit the 15μm CO_2 absorption band to measure radiation emitted from atmospheric levels, so a vertical profile of a structured atmosphere is obtained. Meteosat Second Generation (MSG) carry the Spinning Enhanced Visible and InfraRed Imager (SEVIRI) instrument 0.4–1.6m (four visible/NIR channels), 3.9–13.4μm (eight IR channels), and Geostationary Earth Radiation Budget (GERB) short-wave instrument: 0.32–4μm and long-wave: 4–40μm bands.

The principle of satellite depth sounding is that gas molecules emit radiation at the frequency at which they absorb. Absorption by a gas – for example, CO_2 – is due to different molecular vibrational modes, which occur in a series of bands across the spectrum. Radiometric atmospheric soundings were pioneered by sensors on the Nimbus NOAA satellites. Broad-band spectroscopy sensors such as Nimbus's InfraRed Interferometer Spectrometer (IRIS) retrieve the thermal emission spectra for temperature, humidity and ozone concentration profiles. IRIS measured Earth planetary radiation in broad bands: 6.25–22.5μm. Target radiation is split into equal beams to create an interferogram transmitted to ground. Water vapour, ozone and CO_2 absorption bands produce thermal emission spectra, giving vertical gas concentration profiles. IRIS can probe the atmospheres of other planets, such as Mars, while instruments used on distant planetary missions may likewise be used on terrestrial missions.

5.3 Atmospheric transmission

Atmospheric transmission inversely follows the absorption spectrum. Atmospheric scattering makes a big difference, especially in hazy or foggy conditions. Scattering generally has a larger impact on spectral transmission at sea level than absorption. Viewing angle is critical, with transmission conditions best for vertical paths with the shortest atmospheric path and deteriorating as the slant path increases. As a thicker and longer atmospheric pass is presented, loss increases. This of course suits satellite observation looking earthward, that is, down.

5.4 Radiation from Earth

At wavelengths generally greater than one metre, very little absorption takes place except under extreme weather conditions. However, the most valued IR for Earth observation and emission are the MIR and FIR bands. The main MIR advantage is its sharply defined transmission. The benefit of the FIR is that it lies in the middle of Earth's thermal emission spectrum, with peak IR emission at 10μm. Unfortunately, there is an atmospheric absorption close to 9.6μm. Both windows are used for studies of Earth's surface and cloud cover altitudes by weather satellites such as US Nimbus and Russian Cosmos and Meteor families. The Advanced Very High Resolution Radiometer (AVHRR) system is found on current Tiros-N weather satellites, which are third-generation polar-orbiting environmental spacecraft, a co-operative effort between the USA, France and UK. AVHRR sun-synchronous satellites provide 1.1km high-resolution imagery swaths. The Temperature Humidity Infrared Radiometer (THIR) on Nimbus 5 was designed for temperature evaluation of radiating Earth surfaces in the 10–12μm band IR window and assessment of atmospheric moisture in water vapour absorption at 6.7μm.

In the two main IR windows, clouds can easily obscure Earth's surface. Water vapour is transparent to radiation but water droplets reflect and scatter. Attenuating effects are often *so strong* that even shallow clouds can reduce transmittance (optical or otherwise) to nearly zero. Where clouds are present in an atmospheric column, their upper surfaces act as radiating surfaces as far as sensors designed to exploit atmospheric windows are concerned. A key benefit is that in cloudy areas, equivalent black body temperatures derived from recorded radiation levels can map cloud heights as cloud top temperatures usually decline up through the troposphere. The Surface Composition Mapping Radiometer (SCMR) on Nimbus 5 measured radiation in regions: 8.4–9.5μm and 10.2–11.4μm. Both bands sit comfortably in the wide FIR window 8–14μm. Simultaneous data from both channels yields different equivalent cloud temperatures and it is common for satellite cloud thermography to display 'black hot' surface imagery against a cold 'white' background sky, to improve image contrast.

The microwave region is frequently exploited. Opaque regions arise due to water vapour and atmospheric oxygen. Microwave atmospheric windows have clear advantages over infrared windows, which are used if weather conditions are severe. Long microwave wavelengths, 1mm–1m, penetrate extremely thick cloud and even rain storms. Microwave sensing, whether active or passive, has a high degree of all-weather capability that other spectral systems lack. At long wavelengths scatter is relatively insignificant, and under normal atmospheric conditions absorption can be considered as attenuation with little error.

5.5 Atmospheric composition

At sea level, the main constituents of Earth's dry atmosphere by volume are: nitrogen 78.08%, oxygen 20.95%, inert argon 0.93% and CO_2 0.04%, with traces of other gases: neon, helium, methane, krypton, carbon monoxide, sulphur dioxide, hydrogen, ozone, nitrous oxide, xenon and nitric oxide. There is also a significant but variable amount of water vapour, usually specified as relative humidity H (typically 0.1–0.3%) defined by:

$$H = \frac{p_{water}}{p_{sat}(T)} \quad \text{(eq 5.1)}$$

as a fraction between 0 and 1 (or a percentage), where p_{water} is the partial pressure, defined as the product of total atmospheric pressure and water vapour volume fraction, and $p_{sat}(T)$, which is the water saturated vapour pressure at temperature T.

Example 5.1: At 23°C, if p_{sat} = 2.46kPa and total atmospheric pressure is 101kPa, if the relative humidity H = 0.3, what is the water vapour volume fraction (2 decimal places)?

$p_{water} = p_{partial}$ = total pressure × water vapour volume fraction

$$H = \frac{p_{water}}{p_{sat}(T)} = \frac{101 \times \text{water vapour volume fraction kPa}}{2.46 \text{kPa}} = 0.3$$

So water vapour volume fraction = 7.3×10^{-3}

The atmosphere also contains liquids and solids (clouds or precipitation), dust and aerosol particles, with variable particulate concentration: sand, volcanic ash, insect biomass, forest fire particulates etc. Atmospheric pressure and density diminish with increased height. Molecules affected by gravity sink towards the surface, but cannot because of thermal (Brownian) motion and gases beneath. The height distribution of molecules is closely exponential. However, there are large variations from this overall dependence and it is usual to divide the atmosphere into distinct layers: the **troposphere**, the **stratosphere**, the **mesosphere** and the **thermosphere**, figure 5.3. The mesosphere and thermosphere are often taken together and termed the **ionosphere**. All these layers will be discussed further in Chapter 8. The ionosphere is composed of ionised layers of stratified gases, extending 50–400km above Earth's surface. Ionisation is produced by extreme UV and X-ray solar radiation, which is enhanced when solar Coronal Mass Ejections (CMEs) occur. The ionosphere is extremely unstable and varies spatially and temporally, with electron density variations caused by changes in geomagnetic latitude, time of day, especially within an hour of dusk or dawn, and year (primarily correlated with observed sunspot number). For frequencies below 9MHz the ionosphere is opaque, setting a lower frequency limit for practical satellite remote sensing.

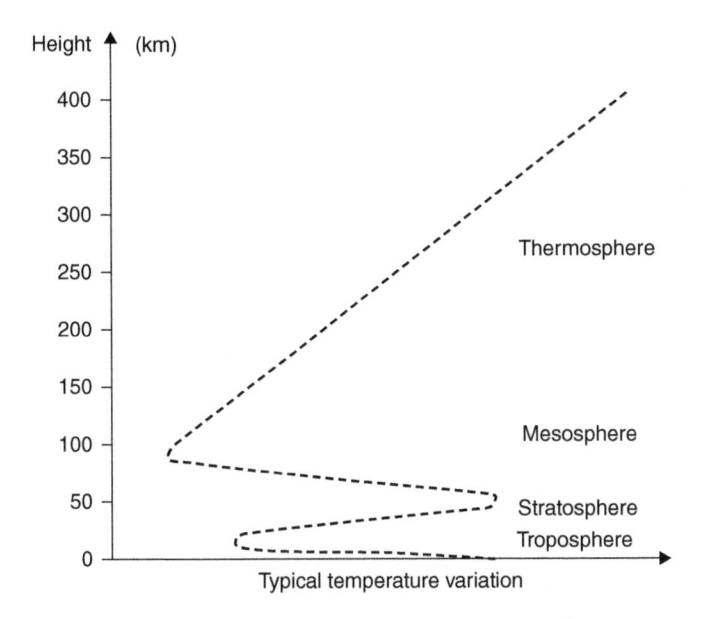

Figure 5.3: *Temperature variation with height through Earth's atmosphere.*

Satellite sensors largely look through the atmosphere straight down, with roughly 90 per cent of the atmospheric mass below 16km altitude. Given that atmospheric pressure and density both diminish with height above Earth's surface, let us consider the pressure recorded at the ground as p_0, and let that at height h be p. If ρ is the atmospheric density, and there is a small decrease in pressure dp for a small increase in height dh, then

$dp = -g\rho dh$ **(eq 5.2)**

Now consider one mole of gas, where

$pV = RT$ (**5.3**), given density $\rho = Mass/Volume = M/V$ **(eq 5.3)**

is written as:

$pM/\rho = RT$ so that: $\rho = pM/RT$ **(eq 5.4)**

Consequently, a small pressure change (5.2) is now given by:

$$dp = -\frac{\rho gM}{RT}dh \quad \textbf{(eq 5.5)}$$

and gathering similar terms together:

$$\frac{dp}{p} = -\frac{gM}{RT}dh \quad \textbf{(eq 5.6)}$$

Therefore integrating w.r.t. p: $\ln p = -\frac{gM}{RT}h + \text{constant}$

Now since $p = p_0$ when $h = 0$, then constant $= \ln p_0$ whereupon:

$$\ln p = -\frac{gM}{RT}h + \ln p_0$$

$$\ln p - \ln p_0 = -\frac{gM}{RT}h$$

$$\ln \frac{p}{p_0} = -\frac{gM}{RT}h$$

And taking inverse logs: $\frac{p}{p_0} = e^{-\frac{gM}{RT}h}$ leading to:

$$p = p_0 e^{-\frac{gM}{RT}h} \quad \textbf{(eq 5.7)}$$

We can substitute the average value for atmospheric molecular density $M = 0.028944$ kg/mol, with $g = 9.8 ms^{-2}$, for a standard terrestrial temperature of $T = 288K$ (15°C) and gas constant $R = 8.314$ J/mol/K with pressure at sea level being p_0. At 10km altitude with the same terrestrial conditions we find pressure of approximately $0.3\, p_0$. Standard atmosphere variations of temperature, pressure and density take place at sea level and as a function of height. Equation (5.7) shows that pressure falls to half its initial sea level value at 5.5km height.

5.6 Atmospheric and ionospheric turbulence

Atmospheric turbulence is always present in the lower atmosphere, causing continual variations in density and air refractive index, mixing gas composition, and preventing a dangerous build-up of increased oxygen concentration at sea level! The *phase* of electromagnetic signals is further corrupted by such variations (such as twinkling lights at night), which adversely influence imaging system behaviour. Lower atmosphere turbulence causes similar effects at radio frequencies, because of the similar refractive index of air at all wavelengths. However, for satellite observations at radio frequencies, the ionosphere provides a serious problem, and can significantly affect satellite GPS signals.

5.7 Cloud, rain and snow

At any time, approximately half of Earth's different surfaces are viewed through cloud cover. Visible sensors, and to a lesser extent those in other spectral bands, are limited by significant cloud cover, which is a problem for satellites that revisit specific locations rarely. It is estimated that Landsat satellites revisit some UK locations (50°–60°N) once every 16 days, and obtain a cloud-free scene in some places only once a year. The probability of seeing less than 10 per cent cloud cover in a single USA Landsat view is 0.05–0.4, needing 5–100 pictures for a 75 per cent probability of less than 10 per cent cloud cover. In 1987, of 200,000 SPOT-1 satellite scenes acquired, only 24 per cent had less than 1.6 okta of cloud, where 1 okta means one-eighth of sky is obscured by cloud.

5.8 Radiation propagation

Scattering In practice, complications arise with Earth's atmosphere lying between a sensor and its observed target. Although the speed of electromagnetic waves is barely affected by the atmosphere, the medium can affect energy propagation characteristics, including:

(i)   propagation direction

(ii)   radiation intensity

(iii) radiation wavelength received by a target at the atmosphere's base

(iv) the radiant energy spectral distribution.

Frequently, radiation direction and intensity are altered by particles of suspended matter in the atmosphere, which redirects radiation en route through a turbid or scattering medium. A proficient understanding of radiation scatter is needed for selection of sensors or filters, and where image degradation due to atmospheric impurities must be avoided, or reduced, by sensing at the best wavelength.

Attenuation Loss resulting from particle scattering suspended in the atmosphere is related to wavelength, concentration and particle diameter, atmospheric optical density and absorptivity. Common scatter sources include atmospheric dust, ash, smoke and haze, hailstones and snow crystals (figure 5.4).

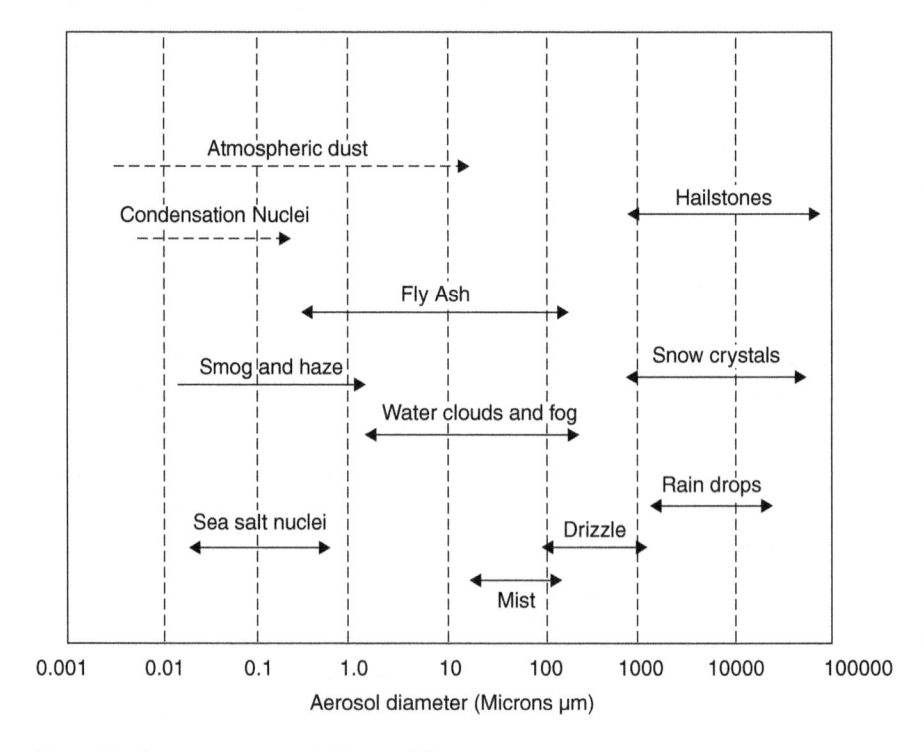

Figure 5.4: Common scatterers and aerosol diameters.

Rayleigh scatter This involves molecules and tiny particles with diameters **much less** than a wavelength with an inverse 4th power dependence on wavelength. For example, UV radiation is scattered 16 times more than red, which explains the orange and red wavelength dominance at sunset when the sun is low in the sky.

Mie scatter This occurs when the atmosphere contains essentially spherical particles with diameters approximate to the wavelength. Water vapour and particles of dust are the main agents when scattering visible light.

Non-selective scatter Particles bigger than the wavelength are involved – for example, water droplets with diameters 5–100μm scatter all visible wavelengths 0.4–0.7μm with equal efficiency. Consequently, clouds and fog appear white.

5.9 Absorbance and transmittance

For a medium radiant energy, transmission is inversely related to the product of layer thickness and extinction coefficient. Transmittance decreases as the combined effects of absorption and scattering increase. The radiation percentage absorbed or **absorbance a(λ)** was introduced in Chapter 1, and depends on atmospheric composition, thickness and wavelength. Absorbance helps choose the best sensing wavelength and to correctly interpret remote sensing surveys. The total power absorbed at a wavelength is found by multiplying absorbance a as a function of wavelength by the spectral radiant flux, that is = **a(λ) × Φ(λ)**. Most absorbed energy is converted into heat so temperature increases make objects into secondary radiation sources, emitting absorbed energy. If we remotely sense targets through suitable atmospheric IR 'windows' and under cloud-free conditions, more is learned than with conventional photography.

Transmittance τ of the atmosphere is defined as the ratio of radiation at distance *x* within the medium over a surface to that incident, which is a function of wavelength, **τ(λ)**. The total power transmitted at a wavelength is given by multiplying transmittance τ as a function of wavelength by the spectral radiant flux = **τ(λ) × Φ(λ)**. There are various molecular absorption mechanisms, such as electronic transitions, vibration and rotation, which can combine. They are covered in detail elsewhere [5.1].

5.10 Ocean attenuation

The impact of the atmosphere on solar output is an example of a modulation transfer function. Similarly, the ocean modulates solar output reaching the surface, and this has an impact on satellite imagery taken over the ocean, particularly images taken over shallow littoral waters. In terms of underwater light attenuation, light levels attenuate with depth rapidly in clear water; blue-green levels reduce to just 1 per cent of the near surface value in 100m of deep water, while red falls more rapidly, dropping to 1 per cent in only 10m (figure 1.11). Intensity falls with depth across the visible and NIR.

Similarly, in water, attenuation is composed of absorption and scattering. Absorption is a mechanism by which energy is lost and converted into heating the water. In scattering, variations in water density cause electromagnetic waves to change their propagation direction from an otherwise 'straight' path – for example, the visible effect observed while walking in a blizzard. In muddy rivers, visibility can be only 0.1m, so a diver has difficulty seeing a hand in front of his face even if just submerged. Deeper in the sea, less light penetrates until finally vision is impossible. Conditions for vision underwater on a good day are comparable with foggy conditions at sea in the above water environment. It isn't the low light intensity that prevents distant viewing above or below the waterline, but rather light scattering from water droplets in fog, which degrade contrast below the minimum required for identifying an object against its background. This is unsurprising as a medium can be translucent (light can pass through), but **not** transparent (clear enough to see images). Absorption is selective and varies in amount and spectral distribution across the oceans, depending on 'water quality'. Water molecules have broad absorption resonances in the visible and IR bands. However, selective absorption also occurs. Under good environmental conditions, dependent upon scattering and absorption, underwater imaging, exploration and surveying of wrecks is possible.

When a photon interacts with a water molecule it may be absorbed, with energy transferred to the molecule. Light absorption is quantised so energy is only absorbed in discrete steps or 'quanta' by atoms or molecules. Electrons have precise energy levels so only certain 'orbits' are permitted. Atoms absorb discretely if sufficient energy is available to raise an electron to the next available orbital, and an energy quantum is absorbed. A molecule has many electronic energy levels associated with its component atoms, besides quantised vibrational and rotational energy levels. Water visible light absorption involves low energy molecular vibration in the visible, rather than high energy electronic changes. Water selectively absorbs low energy (long wavelength red). Consequently, the low absorption of higher energy (short wavelength blue) helps generate water's characteristic blue colour. The absorption coefficient $A(\lambda)$ values for pure water in the visible are given in table 5.1, with $B(\lambda)$, the scattering coefficient, and $C(\lambda)$, the total resulting loss coefficient, accounting for both contributions.

λ nm	Colour	A m⁻¹	B m⁻¹	C = (A + B) m⁻¹
410	Violet	0.016	0.007	0.023
470	Blue	0.016	0.004	0.020
535	Green	0.053	0.002	0.055
555	Yellow-Green	0.06	0.002	0.069
575	Yellow	0.094	0.002	0.096
600	Orange	0.244	0.001	0.245

Table 5.1: *Absorption and scattering coefficients as a function of wavelength.*

Further details about underwater modulation of solar output are found elsewhere [5.2].

Two important multi-sensor satellite atmospheric observation platforms employed were the two European Remote Sensing (ERS) satellites, namely ERS-1 and ERS-2. ERS-2 was launched in April 1995, carrying equipment such as the Global Ozone Monitoring Experiment (GOME). Nearing the end of ERS-1's lifetime, and during initial ERS-2 operations, both satellites operated in tandem for several months. The advantage of this was that two data sets could be compared and contrasted. GOME on ERS-2 was a nadir-looking scanning instrument measuring solar radiation transmitted through, or scattered by, the atmosphere. GOME's main objective was measuring atmospheric ozone. Stratospheric ozone, as opposed to ozone at ground level where it is regarded as a pollutant, plays a key role in survival of life on Earth because it absorbs energetic solar ultraviolet, which is biologically damaging. Stratospheric ozone helps regulate atmospheric temperature and influences global circulation patterns.

In recent decades, evidence has shown that human activities have had a destructive influence on Earth's upper atmospheric ozone levels. Certain chemicals, such as chlorofluorocarbons (CFCs) used as a propulsion gas in aerosol cans and as a coolant within refrigerators, can escape into the atmosphere, where they act as efficient and long-lived catalysts for chemical destruction of ozone. While the international community has recognised the problem and taken steps to reduce ozone-depleting gases entering the atmosphere, ozone loss continues. Data from sensors such as GOME help map seasonal Antarctic ozone 'holes', which develop during the southern hemisphere spring, and similar reduced levels are observed over the Arctic, but to a lesser extent. Long-term ozone monitoring provides information to understand the processes involved and identify long-term trends.

5.11 The remote sensing inverse problem

What is the inverse problem? The problem with remote Earth observation is that we are trying to determine Earth's surface properties *after* radiation has already transited the atmosphere. Clearly Earth's atmosphere has a modifying effect on what we see returned from the target site, combined with actual surface response or emissions. Hence, what we detect is **not** simply the target properties. What we record is the *inverse function* $I = f^{-1}(S)$ and this is used to calculate the properties of the real surface. The direct problem is very much like observation of animal tracks leading away from a hen house (S). The inverse problem helps us describe the animal – the fox that left the tracks – from the tracks themselves, that is, $I = f^{-1}(S)$! Remote sensing requires solution of this so-called inverse problem, using retrieval or inversion theory. The direct measurement, or forward problem, is where a detector measures a signal $S = f(T)$ generated by interaction of radiation with the target surface, atmosphere or clouds and so on to provide the surface properties (figure 5.5).

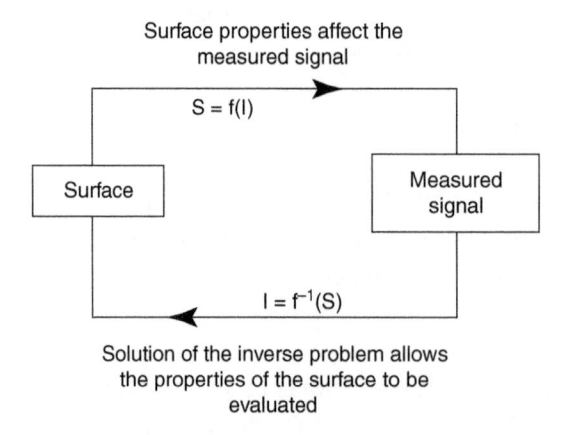

Figure 5.5: *Solution of the inverse problem.*

A typical imaging-based methodology to solve the inverse problem takes the satellite sensor's Digital Number (DN) recorded value, converts the DN value to at-sensor spectral radiance on the satellite, and then converts the spectral radiance to the satellite's at-sensor apparent reflectance [5.3]. Finally, the atmospheric effects due to scattering and absorption are removed, which is the atmospheric correction. With this procedure, we obtain the pixel reflectance at Earth's surface *prior* to the atmospheric effects, with an appropriate radiometric correction. In many cases, satellite imagery provided to the end user is geometrically rectified. One useful tool available to remove atmospheric effects is the Second Simulation of a Satellite Signal in the Solar Spectrum-Vector method, the **6SV** method [5.4–5.5], which yields the reflectance of pixels at Earth's surface – the reflectance value we want. The 6SV code enables accurate simulations of satellite observations, accounting for elevated targets, use of anisotropic and Lambertian surfaces, and calculation of gaseous absorption. The code is based on the method of successive orders of scattering approximations, and is not always a simple process as problems such as non-unique solutions of the inverse problem, or the fact that we obtain *discrete* measurements while the function we are trying to solve is *continuous*, and measurement errors, as always, exist.

Questions

5.1 Discuss what effects refraction, absorption and scattering have on waves travelling between two different atmospheric layers. What can happen to waves on reflection at Earth's surface? Consider the particle *size* in the atmosphere.

5.2 At 26°C, if $p_{sat} = 2.57$ kPa, total atmospheric pressure $= 101$ kPa and relative humidity $H = 0.2$, what is the water vapour volume fraction (3 decimal places)?

5.3 ρ is the atmospheric density, and there is a small negative change in pressure dp for a small increase in height h. Show from first principles how the equation $p = p_0 e^{-\frac{gM}{RT}h}$ is derived.

5.4 Using the equation $p = p_0 e^{-\frac{gM}{RT}h}$, substitute the average atmospheric molecular density value $M = 0.028944$ kg/mol, $g = 9.81$ ms^{-2}, for a standard terrestrial temperature of $T = 288$ K, and gas constant $R = 8.314$ J/mol/K, with sea level pressure being $p_0 = 101325$ Pa, to find the pressure at a height of 8km in Pa (3 significant figures).

5.5 Taking the values of g, M, R and T to be those in question 5.4, use equation (5.7) to estimate the pressure change for a height decrease at the bottom of the Dead Sea, 430.5m below sea level. Actual recorded values are 106100–106500 Pa. Suggest reasons for this difference.

5.6 Consider the scattering coefficient at 600nm $B = 0.001$ m^{-1} and the scattering coefficient at 400nm $B = 0.007$ m^{-1}. If the scattering coefficient falls off as the inverse 4th power with wavelength, estimate the relative difference for the 400nm to

600nm scattering coefficients (3 decimal places). Can you suggest reasons why *predicted* and *actual* values are different?

5.7 An increase in air pressure of 1hPa lowers water level by about 1cm. Explain how *higher* air pressure causes lower sea levels.

5.8 Explain how remote sensing helps weather forecasting and to improve forecasting *accuracy*.

5.9 Explain what the remote sensing inverse problem is and how it can be solved using the Digital Numbers recorded at individual sensor pixels.

5.10 Using the internet, explain the importance of ocean colour to MODIS and SeaWiFS. How are space-based sensors such as colour scanners, multi- and hyperspectral systems, SAR, microwave and IR radiometers, altimetry and scatterometer data useful for water quality pollution monitoring?

References

[5.1] *Handbook of High-resolution Spectroscopy*, Martin Quack and Frederic Merkt (Eds), (John Wiley & Sons, Inc., New Jersey, 2011, ISBN 9780470749593).

[5.2] *Essential Sensing and Telecommunications for Marine Engineering Applications*, Christopher Lavers (Bloomsbury Publishing, London, 2017, ISBN 1472922182).

[5.3] 'GeoEye-1 Radiance at Aperture and Planetary Reflectance', NE Podger et al., 11 April 2011, accessed 21 March 2017, apollomapping.com/wp-content/user_uploads/2011/09/GeoEye1_Radiance_at_Aperture.pdf

[5.4] '6SV Routine', accessed 21 March 2017 6s.ltdri.org/

[5.5] 'Validation of a vector version of the 6S radiative transfer code for atmospheric correction of satellite data. Part I: Path radiance', SY Kotchenova et al., *Applied Optics*, Vol. 45, No. 26. (2006), pp.6762–6774.

6
Hydrosphere and Cryosphere Applications

'The sea is the universal sewer where all kinds of pollution end up conveyed by rain from the atmosphere and from the mainland.'

Captain Jacques Cousteau, Proceedings before the Committee on Science and Astronautics, US House of Representatives, Jan. 26–28, 1971. No. 1, 325.

6.1 Water resource applications: the hydrosphere and the cryosphere

In this chapter we combine liquid water and frozen water applications, where possible in sequence of visible, thermal and microwave study. In reality, satellite and remote sensing users look at solving particular Earth observation *problems* and use all tools at their disposal to deal with that application. Hence, the focus of these later chapters is on applications, and *not* spectral wavelengths taken in isolation. This is an important point and should be grasped at the earliest opportunity. Satellite imaging provides data for both water and land environments, as they all, unless in geostationary orbit, pass over both surfaces; for example, Aqua and Terra are satellites 'dedicated' to water- and land-based monitoring respectively but contain similar sensors providing data over both, important in water resource applications such as pollution detection, lake shrinkage, wetland mapping, sea ice and snow monitoring, as well as underwater optics. In Chapter 2 we considered the main differences between fresh and sea water in terms of physical and chemical characteristics, and we will now look at how these apply to marine and wetland systems. In Chapter 7 we will see how to assign water land classification in terms of land use. Generally, all three spectral regions – visible, thermal and microwave – provide complementary information. We won't show imagery for every band discussed, but a cross-section of typical imagery for them.

The hydrosphere and cryosphere are critically important to many nations, especially those with extensive seaboards (over 150 plus UN members), and given that seas cover two-thirds of the world's surface. Over 70 per cent of humanity lives within 100 miles of the coast, where almost all major cities and civilisation centres are found. Global economy depends primarily on maritime trade; 99.5 per cent of global trans-oceanic trade is seaborne, accounting for 80 plus per cent of total global trade. In 2016, more than 10 billion tonnes was carried (source: UNCTAD), compared with only 2.5 billion in 1971, and this figure is estimated to grow to 23 billion tonnes by 2060. Many resources are 'carried' by sea: fish, oil and gas (almost 50 per cent), minerals, renewable energy: wind/wave and tidal, pipelines and critical telecommunications cables (carrying 98 plus per cent of global internet traffic). Clearly, changes in sea level, sea disturbances or maritime power projection have potential to influence the bulk of the world's population. Consider the UK, an island state with 10,000 plus miles of coastline, about 600 ports, nearly 300 offshore oil and gas platforms, with growing dependence on power generated offshore by renewable energy wave hubs: www.wavehub.co.uk/, with an Exclusive Economic Zone (EEZ) of 2.5 million miles^2, the fifth largest worldwide.

6.2 Water pollution detection

All natural water contains impurities, and is considered polluted if impurities limit water use for domestic or industrial purposes. Not all pollutants result from human activity; some natural pollution sources include soil-leached minerals and decayed vegetation. There are two source types: point and non-point. *Point* sources are highly localised, such as industrial outfalls. *Non-point* sources, such as fertiliser and sediment run-off from agricultural fields, have large and dispersed source areas and are hard to locate exactly.

Each of the following materials categories, when present in excess amounts, results in water pollution:

(1) Organic wastes from domestic sewage and industrial wastes of plant and animal origin remove oxygen from water through decomposition.

(2) Infectious agents in domestic sewage and industrial wastes transmit disease, such as *Escherichia coli*, typhoid, cholera, dysentery etc.

(3) Plant nutrients (fertilisers) promote excessive and rapid growths of aquatic plants such as algae or underwater weeds.

(4) Synthetic organic chemicals, such as detergents and pesticides, are toxic to aquatic life, and potentially humans.

(5) Inorganic chemical and mineral substances from mining etc. can pollute naturally pure rivers, destroying fish, or enter the food chain and impact humans – for example, methyl mercury in Minamata, Japan.

(6) Sediments fill up streams and harbours etc.

(7) Radioactive pollutants from mining or nuclear testing etc.

(8) Temperature increases from water used for cooling power plants. When Chernobyl Nuclear Reactor 4 exploded in 1986, Landsat 5 was the first civilian satellite to confirm the disaster. Images acquired before the explosion show heated water pumped from the plant into the adjacent cooling lake, figure 6.1 (left). Under normal circumstances, Landsat 5's IR band 6 would show heated water orange, or blue as it cools. The 29 April image shows all the lake water at the same low temperature, evidence the reactor stopped operating normally, figure 6.1 (see plate section).

Sediment pollution is often depicted on satellite and aerial photos as water plumes dispersing from silt-laden rivers. Often in the immediate period before seeing a plume, rain falls on the river's watershed. Soil erosion from agricultural fields, coupled with heavy stream flow, may result in transport of large volumes of suspended silt and clay. Clear lake water has low sunlight reflectance, but because the spectral response of suspended materials is distinct from natural lake water, materials are easily distinguished on imagery [6.1].

6.3 Lake eutrophication

Inland lake water quality is described by its **trophic** nutritional state. A lake choked with algal blooms or aquatic weeds is a eutrophic nutrient-rich lake. Lake ageing, or **eutrophication**, is a natural process, but anthropogenic activities may accelerate and result in polluted water. Such processes relate to land use/cover. Floating algae is a good indicator of lake trophic status. Excess blue-green algae are present under eutrophic conditions. Seasonal blue-green algae blooms occur under warm late-summer conditions, with green algae present throughout a lake's annual cycle. Narrow-band multiband selective filter spectral response photography maximises detected differences in water body chlorophyll concentration. With increased concentration, green reflectance increases significantly and blue decreases. Water body colour due to algal blooms is covered in section 6.6.

6.4 Ice shelves *visible*

Ice shelves surround three-quarters of Antarctica's coastline. Floating glacial ice shelves play a role in stabilising Antarctica's ice mass balance. Ice shelves grow and shrink, gaining mass from glaciers that flow into them over land, from snow and from sea water freezing beneath. They lose mass by calving icebergs or melting below. Icebergs regularly break off or collapse from ice shelves, and this can happen quickly. A large portion of the Larsen B ice shelf, located on the Antarctic Peninsula, collapsed in 2002. Over the course of a month, 3250km^2 of shelf collapsed.

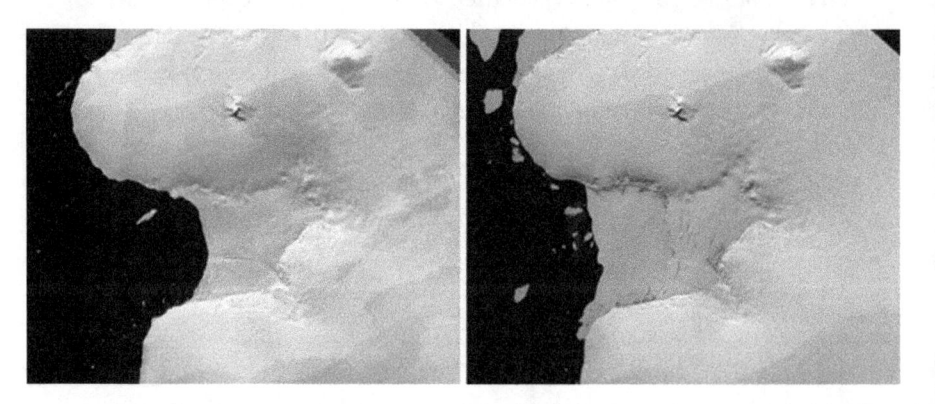

Figure 6.2: (left): *15 Feb 2000 Landsat 7 Verdi Ice Shelf;* (right): *27 Dec 2013 Landsat 8.*

Shelf collapse can make glaciers that flow into them more unstable, flowing and receding faster. As glaciers accelerate and recede, more ice ends up in the ocean, contributing to sea level rise (figure 6.2).

6.5 Water security issues

6.5.1 Aral Sea case study – optical

The Aral Sea, Kazakhstan, was once the fourth largest lake in the world. In the early 1960s the Soviet Union undertook an ambitious irrigation project, diverting the Amudarya and Syrdarya rivers, which fed the Aral Sea, to create a huge cotton belt. Since then the Aral Sea has turned to desert. It shrunk to under half its former size, and water salinity increased fourfold (figure 6.3). Drinking water quality was severely affected by soil salinisation and inadequate land drainage. Rivers are now polluted with toxic chemicals from heavily fertilised cotton fields. Pesticide and heavy metal traces are found in mothers' milk, with infant mortality high and congenital deformations common. There is a high incidence of thyroid and oesophageal cancers and of gastrointestinal and respiratory problems. Difficult living conditions and unemployment caused by a declining fishing industry forced people to move away, although there has been some reversal of lake shrinkage with the construction of the Kokaral Dam across a narrow stretch of the Aral Sea, splitting off the North Aral Sea from the larger South Aral Sea. The dam is conserving the dwindling waters of the Syrdarya river and maintaining, and hopefully reviving, the damaged ecology. A complete time sequence between 2000 and 2017 is visible on the NASA website [6.2].

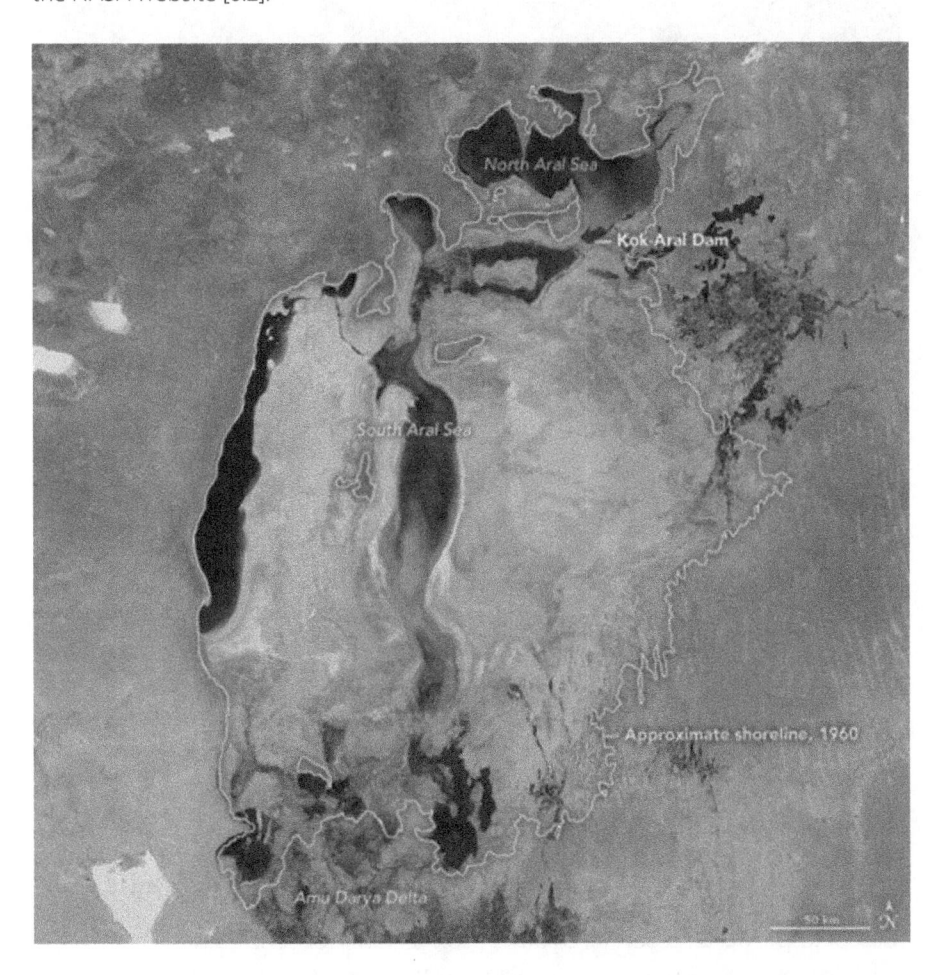

Figure 6.3: *The original extent outline of the Aral Sea, with the current visible water, 2017. Courtesy NASA.*

Similarly, we can monitor surface water changes of other receding lakes, such as Lake Chad, or changes in lake quality and biodiversity, such as Lake Victoria. Industrial use of salt and mineral products in lakes such as the Dead Sea (figure 6.4) can transform a region. The Jordan River is the only water source flowing into the Dead Sea, although there are small springs under and around it, forming pools and quicksand pits along its edges. Rainfall is 100mm per year towards the north of the Dead Sea and 50mm in the south. The Dead Sea zone's aridity is due to the rain shadow effect of the Judean Mountains.

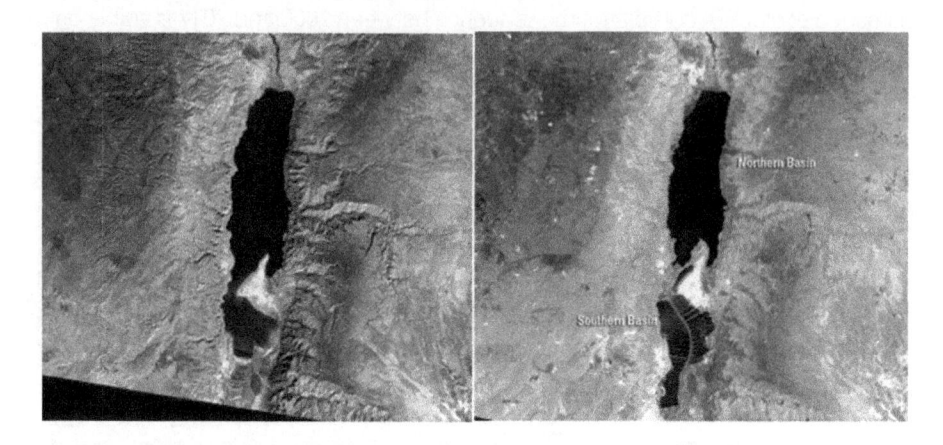

Figure 6.4: (left): *Dead Sea, 1 Jan 1973, Landsat 1;* (right): *13 June 2017, Landsat 8, NASA.*

Photo interpretation helps identify aquatic vegetation, particularly within the IR bands. Groundwater location is important for water supply and pollution control analysis in regions with water security issues, such as the Jordan river between Israel and Jordan.

6.6 Ocean colour *visible*

Ocean colour is a measure of biological activity, such as phytoplankton levels, and particulate matter. It measures biological productivity, marine optical properties, interaction of winds and currents with ocean biology, and human influences on the marine environment. Phytoplankton is the main primary producer at the base of the maritime food chain, microscopic green photosynthesising 'factories' floating in the sunlit upper levels. Generally with ocean colour, Earth observation uses multispectral data. Several colour sensors are available. The Coastal Zone Color Scanner (CZCS) provided data between 1978 and 1986 in 1 and 4km grid resolutions, with data products available [6.3]; SeaWiFS data is available since 1997, with 1 and 4km grid data, with products available for scientific use [6.4]; and MODIS (Moderate Resolution Imaging Spectroradiometer) is available since 2000 with 1km grid data, and products freely available [6.5].

6.6.1 Optical chlorophyll detection of phytoplankton blooms with SeaWiFS

As photosynthesis takes place in chlorophyll-a, aquatic algae blooms impart a green colour to surface water, which is observed by satellite remote sensing. Calibrated ocean optical observation from satellites such as SeaWiFS, launched in 1997, are available. SeaWiFS measured sea algae blooming, and areas of interest on land as well. Satellites performing this task today include Aqua.

Figure 6.5: *An Aqua/MODIS scene of the Gulf of Aden on the west of the Arabian Sea, showing phytoplankton carried by turbulent surface currents. Data collected 12 Feb 2018. © NASA (oceancolor.gsfc.nasa.gov).*

6.7 Ocean wind *microwave*

Ocean wind measurements help us understand and predict severe weather patterns, as winds transfer energy between the atmosphere and the oceans. Surface roughness can be monitored with SAR radar and scatterometer backscattered returns, which correlate with wind speed and wave height (see Chapter 4).

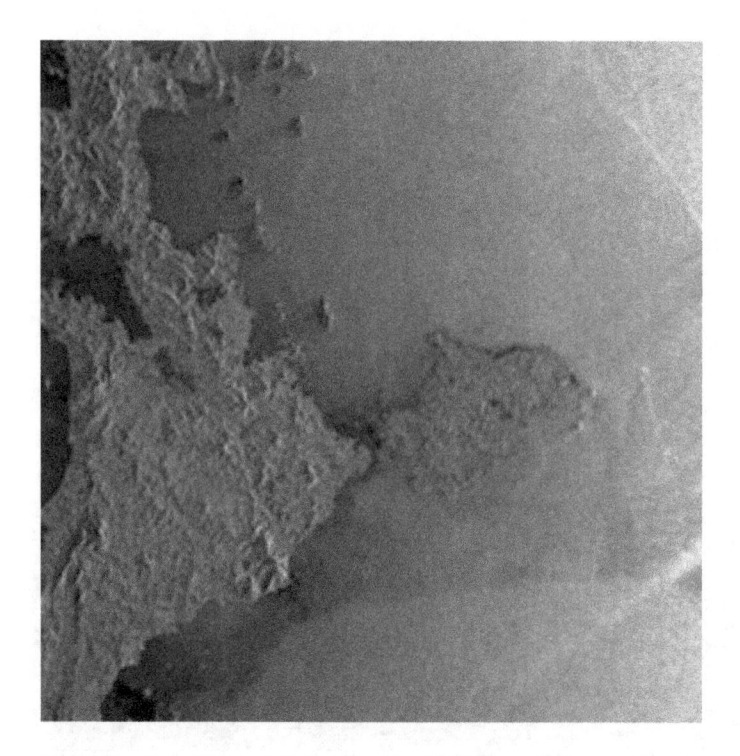

Figure 6.6: *ERS-2 SAR image showing wave direction and height, acquired over the Sulu Sea near Palawan Island (08' N, 119° 10' E, Philippines). Several small-scale rain cells located near an atmospheric front are visible.*

Wind data is provided from the NASA Scatterometer (NSCAT) with data from 1996 to 1997, C-band with 50km grid resolution, and QuikSCAT SeaWinds Scatterometer, launched June 1999, C-band with 25km spatial resolution, for wind vectors 3–20ms^{-1} with 2ms^{-1} accuracy. Data from NSCAT and QuikSCAT are available freely [6.6].

6.8 Rivers

Water resource applications of satellite and aerial photography include: hydrological watershed assessment, reservoir site selection, shoreline erosion, snow cover mapping and recreational lake and river use. For flooding – that is, the extent of surface waters – and sediment concentrations, see MODIS and QuikSCAT data products. NASA's global flood monitoring system, operated through the US Dartmouth Flood Observatory, provides multi-temporal MODIS data. Mapping surface water extent during flooding improves flood plain mapping. One of the best documented flood events is Hurricane Katrina, where winds and storm surge reached the Mississippi coastline on 29 August 2005, beginning two days of destruction through central Mississippi. Many coastal towns were obliterated in a single night. Hurricane-force winds reached the Mississippi coast, lasting 17 hours and generating tornadoes and a 9m storm surge, flooding far inland; 238 people died in Mississippi, but many survived by climbing onto rooftops or swimming to higher buildings and trees. The worst property damage from Katrina occurred in coastal Mississippi, where all towns flooded over 90 per cent and waves destroyed many historic buildings.

Damming of rivers can be investigated using high-resolution satellite imagery. One example relevant to river image analysis in inaccessible regions is the Gilgel Gibe III dam, a massive hydro-electric dam completed on the Omo River, Ethiopia, to support vast commercial plantations, forcing local tribes to leave their land [6.7]. The government is planning to build Gibe IV and V. These dams have potential to destroy a fragile environment and the livelihoods of tribes [6.8], closely linked to the river and its annual flood (figure 6.7).

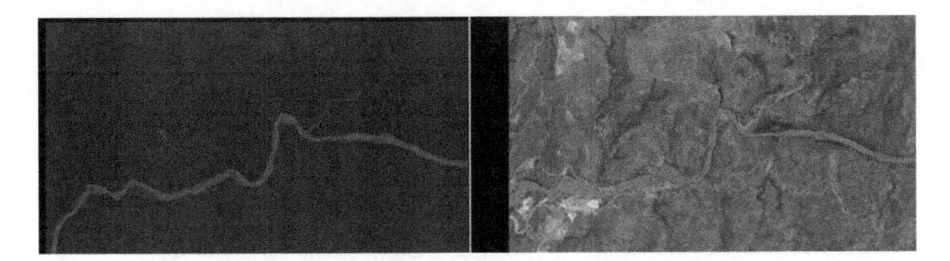

Figure 6.7: *The Omo River before construction of the Gilgel Gibe III dam (left) and during construction (right). Courtesy GeoEye Foundation Imagery Award, and recent publications [6.7–6.8].*

6.9 Wetland mapping

The world's wetland systems contribute to a healthy environment in many ways, retaining water during dry periods and keeping water tables high and relatively stable. During periods of flood, they reduce flood levels and trap suspended solids and attached nutrients. Urbanisation removes wetland systems and causes lake water quality to deteriorate. One notorious drainage of marshes took place in Iraq between the Gulf Wars. The marshes were for some time considered a refuge for elements persecuted by the government of Saddam Hussein. During the 1970s, irrigation project expansion began to disrupt the water flow to the marshes. After the first Gulf War (1991), the Iraqi government aggressively revived a programme to divert the flow of the Tigris and Euphrates rivers from the marshes in retribution for a failed Shia uprising. This was done primarily to eliminate the food sources of the Marsh Arabs and prevent militiamen taking refuge in the marshes, militias having used them as cover. The plan systematically converted wetlands into desert, forcing residents out of their settlements [6.9]. Marsh villages were attacked and burned down and there were reports of water being deliberately poisoned [6.10].

6.10 Surveillance maritime applications

Remote surveillance of sea-based ship movements and target classification is of interest to governmental organisations, and complements remote navigational collision avoidance aids such as S-AIS and GPS to maritime users. We will examine visible, thermal and microwave surveillance applications for both terrestrial (sea/air/land) platforms and space-based satellite platforms. After the Cold War, military visible spy satellites, thermal and radar technologies started to enter the civilian sector, providing benefits to coastguards and navies worldwide. Benefits of all three spectral bands exist to terrestrial (sea/land/air) users, as well as those obtained from space on satellites.

6.10.1 Terrestrial

6.10.1.1 Terrestrial maritime visible imagery

Ship, UAV and aircraft platforms (encompassing maritime surveillance aircraft) are regularly employed for various applications including: ship monitoring to tackle illegal unregulated and unreported fishing, criminal activity including illicit trades such as narcotics, port observation, ice navigation, natural and water resource management, coastal changes, oil spill detection underwater reef mapping, anti-piracy, monitoring Economic Exclusion Zone (EEZ) patrols, as well as long-range search and rescue. Some applications will be explored in terms of non-visible imagery.

6.10.1.2 Terrestrial maritime thermography

Thermography is the remote sensing branch that measures radiant temperature of Earth surface remotely, detecting ships at sea or ocean currents. Terrestrial thermography involves Earth-based sensors operated at close range – for example, a ship-based camera. Space-based sensors are usually operated in Low Earth Orbit (LEO). Aerial thermography involves sensors from aircraft or, increasingly, UAVs, offering the benefits of over-site targeting without satellite launching logistical problems. Thermal imaging is emerging as a valuable resource for applications such as maritime rescue, aircraft maintenance and process monitoring.

Finding missing persons Detecting missing or hiding persons is common. In maritime search and rescue, a person in water at night can be detected easily on account of their radiated thermal energy. The technology is also maturely used by law enforcement agencies at night.

Uncovering the concealed Thermal imaging can see through smoke and dust, secure public buildings and port facilities, and provide surveillance. Thermal imaging highlights people, boats and objects in complete darkness. IR cameras translate heat into high-contrast video pictures, making warm objects show up red against a cooler blue background using *false colour representation*. In total darkness, it is easy to detect illegal immigrants crossing borders by boat or sea or avoiding perimeter patrols. Search and rescue teams in co-operation with firefighters can move safely inside ship compartments or buildings to find people trapped, determine gas or oil leaks, or detect weak spots in dams holding back floodwater. Firefighters can see through smoke to detect hot spots to determine a fire's source, such as in a smoke-filled bulkhead.

Sea safety Public vessel transport, Coastguard, military, pleasure boats and oil rigs manoeuvre at night. Dimly lit docks are hazardous, but with thermal imaging, obstacles, sea ice, ice shelves, buoys and floating debris may be spotted easily at relatively long ranges. Coastguard and military personnel often mount rescue operations in the middle of the night or in foul weather. Detecting heat differences between objects in a 'scene' is crucial in poorly lit dock areas and in boating accidents. The SAFIRE (FLIR) thermal imager family is available in a shipboard version for maritime patrol, navigation assistance and rescue.

Oil platforms are notorious for high maintenance levels and hostile operating environments. Preventive maintenance and non-destructive examination of internal pipe surfaces is reduced with portable IR analysis cameras. Camera systems identify 'hot spots' in worn bearings and components within motors and pumps and can confirm data supplied by diagnostics equipment. Thermal imaging is proven useful in pipe surveys and pressurised vessels susceptible to salt water 'scale' build-up, appearing cooler. British Petroleum (BP) often assesses build-up levels and responds before blockages occur.

SAR maritime monitoring (see Chapter 4) SAR modes include vessel and oil spill detection, sea ice monitoring and iceberg detection. Maritime surveillance radar is discussed elsewhere [6.11], but maritime use is introduced. Surveillance radar are vital to modern naval fleets and their ability to detect small targets in rough seas.

6.10.2 Space-based maritime applications

6.10.2.1 Space-based visible

Many providers of space-based visible maritime Earth observation exist, with the lead provided currently by Digital Globe [6.12]. Applications include: oil spill detection, vessel detection and identification, ice navigation and tracking, monitoring oil and gas activity in the High North and Arctic, unregistered fishing, port harbour, gas and oil platform maritime installation monitoring, disaster and flood mapping, and others listed in 6.10.1.1.

Other optical and multi-sensor systems include NASA's Ice, Cloud and Land Elevation Satellite (ICESat), launched in 2003. Its primary mission is to map ice sheet elevation and changes using the Geoscience Laser Altimeter System (GLAS), with a frequency-doubled laser at 1064nm for altimetry and a 532nm green laser for atmospheric characterisation (aerosols). It produces 40 pulses per second, with a 70m diameter ground footprint, spaced every 175m. Returning photons are collected through a 1m diameter telescope. Return time, power and waveform provide important information about the surface. It is in a near polar orbit, with 94° inclination at 600km altitude.

Poor resolving ability of even relatively close near and MEO space-based thermal systems, such as Landsat 120 × 120m cell resolution, prevents detection of even tanker-sized targets. Weather satellites like MSG provide a global view of weather patterns in relevant bands, but thermal bands can provide oil spill imagery as a result of emissivity differences and thus contrasted radiated signals (figure 6.8).

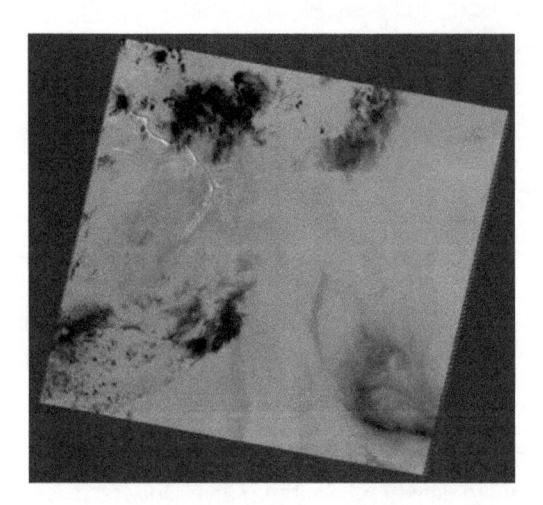

Figure 6.8: *Thermal band 6 TIR oil spill detection. Sentinel-2 data courtesy of Copernicus/ESA, with visualisation by Pixalytics Ltd, notably Dr Sam Lavender.*

6.10.2.2 Space-based radar
Active radar

Again, the role of radar for maritime surveillance involves SAR imagery. SAR provides remote space-based surveillance of sea-based movements and ship target classification. Ship detection is a key role of maritime surveillance as it allows monitoring of maritime traffic, illegal fishing and border activities. With SAR, ships are detected given the distinct radar signature difference between a ship and background sea clutter. SAR satellite imagery provides important details such as ship dimensions, orientation and location. However, satellite SAR imagery on its own has limitations and it is generally combined with AIS data or, increasingly, S-AIS for ocean maritime surveillance. Imaging radar has found wide applications in oceanography. Surface wave fields are imaged distinctly, despite difficulties of deducing wave spectra from SAR images. Coastal feature wave diffraction and refraction by bottom topography variations is often visible (figure 6.9).

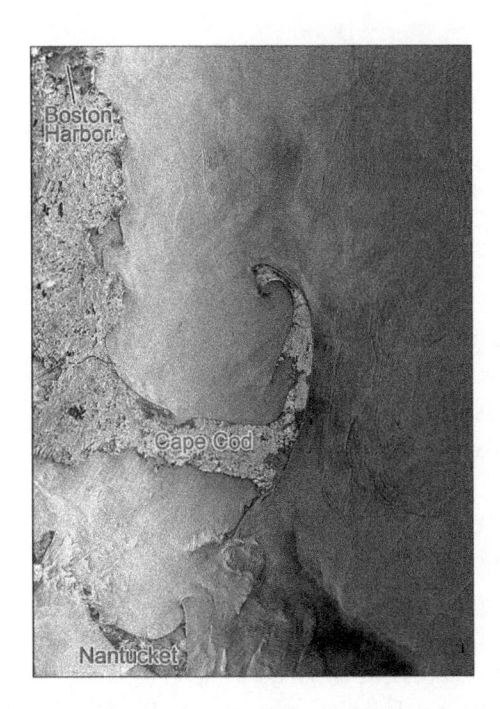

Figure 6.9: *Internal waves and shallow subsea features imaged by SAR Cape Cod, Massachusetts. Both were generated by tidal currents in the region; the image was acquired 27 August 1978 (credit: NASA [6.13]).*

Sea Surface Temperature

Sea Surface Temperature (SST) is of great oceanographic and meteorological importance, particularly events such as El Niño. Thermal sensors provide good spatial resolution and accuracy, with 20 plus years of available data. However, thermal sensors are often obscured by clouds and atmospheric correction is required, and we measure SST to about 10 microns depth, characterising surface water only, which may not indicate water internal bulk temperature. For example, on a day of low humidity, a water body with a high temperature has strong evaporative cooling effects at its surface. Although bulk water body temperature may be warmer than its surface, thermal sensors record **only** surface temperature. SST is deduced from calibrated thermal infrared data, as a pure water surface has an emissivity of 0.993. Atmospheric attenuation may reduce the signal, and is modified by sky reflections. Satellite observations may deviate from calibrations by up to 10K. In Chapter 3, we saw that to combat this, two thermal IR spectral bands are used. So two thermal IR spectral bands have a linear relationship of the form

$$SST(True) = AT_1 + BT_2 + C \quad \textbf{(eq 6.1)},$$

reliable to 0.5K, where T_1 and T_2 are the 'brightness temperatures' measured at two wavelengths, 8 and 10 microns, where A, B and C are constants, providing improved accuracy (see example 3.6, section 3.8.1.)

Various satellites are used, such as AVHRR, MODIS and TRMM (microwave imager), but show spatial coverage differences. Satellite resolution is poorer than that obtained with terrestrial or aerial systems. For example, Landsat resolution is 120 × 120m, but future dedicated NEO satellites (NEOSATs) may provide better resolution. Figure 6.10 (see plate section) shows three views of the same marine surface.

The left image shows icebergs in natural colour. The thermal image, right, shows the same area in false colour. Clouds over the ice shelf don't show up well thermally because they are nearly the same temperature as the shelf. Thermal imagery has an advantage of showing where cold ice ends and warm waters of the Weddell Sea start. It indicates differences in ice type thickness. The mélange is thicker (colder signal), but thinner (warmer signal) than the shelf and icebergs. In several weeks of observations, scientists saw the passage widen between the main iceberg and the shelf (figure 6.11, see plate section).

Sea Surface Height

Sea Surface Height (SSH) changes with wind patterns, ocean currents and ocean temperature (thermal expansion). Radar altimeters measure SSH changes through timing of radar pulses to give the distance between the sensor and the surface. TOPEX/Poseidon, launched in 1992 and still operating, measures SSH between 66°N and 66°S latitude in the C and Ku radar altimeter bands. SSH is accurate to 4.2cm, with data available freely [6.15]. Another satellite, Jason-1, similar to TOPEX/Poseidon, was launched in 2001 and some data is also available [6.16].

Sea snow and ice mapping

Timed series of passive microwave data is available to monitor sea ice extent and is the longest measured remote sensing parameter available. It is hard to distinguish snow from ice in the visible, and NIR using satellite and thermal imaging works well when snow and ice surfaces are visible, although cloud is hard to distinguish from snow when at similar temperatures. However, microwave systems have the advantage that data is collected through clouds and in darkness. Around Antarctica's edge, seas are frozen all year but the *amount* of sea ice increases in winter, extending into the Southern Ocean. To understand Earth's climatic system, it is important to monitor ongoing changes (long- and short-term) in sea ice at Arctic and Antarctic latitudes. Energy exchanged between the oceans and atmosphere significantly affects weather systems as sea ice insulates the sea from a cold atmosphere above it. When sea ice melts or the upper sea surface freezes, it affects upper ocean temperature distribution, which in turn affects ocean current patterns.

This Sentinel-1 radar composite image, figure 6.12 (see plate section), is of the north-east tip of Ellesmere Island (lower left), where the Nares Strait opens into the Lincoln Sea in the Canadian Arctic. The image was created by combining three radar scans from Copernicus Sentinel-1, captured December to February. Each image is assigned a colour – red, green and blue – to create a colour composite. Colours show changes between acquisitions, such as ice movement, while static land masses are grey. The obvious distinction between red and yellow depicts how ice cover changes over the three-month period.

Passive microwave

Passive microwave sensors use the radiometry approach of Chapters 3 and 4. Passive microwave sensors can see through clouds (transparent) and are relatively insensitive to atmospheric effects, but have lower spatial resolution and accuracy than thermal systems. They are sensitive to surface roughness (waves) and precipitation, and can measure SST to a depth of 1mm. Passive microwave radiometer data can provide SST to an accuracy of 1K derived from ATSR data.

ATSR details, carried on SeaSat, Nimbus, ERS-1 and ERS-2 satellites, are introduced in Chapter 9. ATSR satellites occupy near-polar circular orbits at 770km altitude. Passive microwave radiometry is used over ocean surfaces to determine wind speed, to 2ms^{-1} accuracy. The method relies on the influence of wind speed upon surface roughness, and thus emissivity. Salinity is determined through its effect upon emissivity.

ESA's Soil Moisture Ocean Salinity (SMOS) Earth Explorer mission is a radio telescope in orbit, pointing towards Earth. Many visualisations and much data is available [6.17]. Its Microwave Imaging Radiometer uses an Aperture Synthesis (MIRAS) radiometer such as SAR to pick up small microwave emissions from Earth's surface, mapping land soil moisture levels and ocean salinity. These are the key geophysical parameters: soil moisture for hydrology studies and salinity for enhanced understanding of ocean circulation, both vital for climate change models.

Radar images map location of sea ice, as sea water returns a different signal to ice-covered seas. The extent and concentration of sea ice is measured from passive microwave sensors, and data is available [6.18].

For this application, passive microwaves, AVHRR and MODIS are used extensively. Passive microwaves provide typically 25km grid scales, with weekly data since 1978 and daily since 2000. AVHRR provides 1km grid scale data, weekly available since 1996, while MODIS provides 500m grid scale, with data available since 2000. All data is available [6.19]. MODIS data is available daily and 8–10-day composite global snow cover products with 500m spatial resolution [6.20]. There is also data from the National Operational Hydrologic Remote Sensing Center (NOHRSC) providing daily, 1km North American snow cover. This topic is covered further in Chapter 7, with ASTER data mapping glacier extent snow line and glacier topography, while SAR can map glacier zones with interferometry and mapping glacier velocity. Altimetry data related to sea ice and land ice thickness in the Arctic and Antarctic is available from the CryoSat-2 mission [6.21]. CryoSat-2 is an ESA environmental research satellite launched in 2010. It provides data about the polar ice caps and tracks thickness changes in ice with 1.3cm resolution.

6.11 Oil spillages

Materials that form thin films on water surfaces, such as oil, can be detected through satellite or aerial photography. Oil enters the world's water bodies from various sources, including waste discharge and shipping losses. Today, the main industrialised nations obtain most of their energy from petroleum imported via tanker ships from overseas. Oil transportation accidents and run-off account for much of the oil entering the oceans each year. Thick oil slicks have a distinct black or brown colour. Thinner oil has a characteristic silvery sheen or iridescence. Oil films are best viewed between 0.30 and 0.45μm.

Oil is characterised into several categories visually. *Mousse* is a brown emulsion of oil, air and water, forming thick streaks, and looks somewhat like chocolate mousse. *Slick* is a relatively thick layer of a brown or black colour. *Sheen* is a thin silver layer on a water surface, with no black or brown colour. *Rainbow* is very thin iridescent multi-coloured bands visible on water surfaces. Rainbow and Sheen are often lumped together because they are difficult to distinguish. Typically, 90 per cent of spillage volume is concentrated in 10 per cent of the surface area.

6.11.1 UV fluorescence

UV fluorescence is probably the most sensitive remote sensing method available for oil on water monitoring, and can detect films as thin as 0.15μm. UV can stimulate oil to fluoresce, so it is detected on sea water with UV fluorescence, when a powerful visible laser source illuminates the sea. In the absence of oil there is no significant backscattered fluorescence but with oil present there is usually a large backscattered visible fluorescence. Daylight and a clear atmosphere are needed to acquire images, as UV is scattered strongly by the atmosphere. Floating foam and seaweed patches also have bright UV emission, which may be confused with oil. Most UV systems are passive; they rely on sunlight and require little power. Active UV systems were developed for aircraft, with a UV laser irradiating water and stimulating oil to fluoresce. Fluorescence is recorded as a spectrum, **not** an image, and can detect slicks in unexplored basins, helping successful exploration. British Petroleum used an active UV system, the Airborne Laser Fluorosensor (ALF), to explore oil in offshore basins, as accumulations often leak hydrocarbons to form slicks.

Lidar is a common airborne remote sensing technique to produce swaths with millions of data points, so detailed digital terrain models can be mapped and analysed. Lidar surveying can safely collect data in inaccessible coastal areas, including soft cliff and saltmarsh areas, where other survey methods are unsuitable.

6.11.2 Visible and IR oil spillage imagery: Gulf War case study

During the 1999 invasion of Kuwait, Iraqi forces released 4–6 million barrels of crude oil into the Arabian Gulf, covering 1, 200km^2 of water and 500km of Saudi Arabian coastline. This spill was the largest recorded in history. Retreating Iraqi forces also set fire to oil wells, recorded on AVHRR, Meteosat and Landsat satellite imagery and by advancing Allied journalists. Burning oil created giant smoke plumes and spilled further oil into the desert sands (figure 6.13, see plate section).

These images show spillages in visible colour imagery. Other images show visible, reflected NIR and thermal IR bands. Individual TM bands of the Arabian Gulf spill were digitally processed to extract information and create three visible, three reflected NIR and an FIR thermal band image. Reflected NIR bands 4–5 and 7 extract the most information. *National Geographic* Vol.180, No.2, August 1991 provides a good article on the spillage aftermath and its impact on local wildlife and scenery.

6.11.3 Thermal IR oil images

Both oil and water have the same kinetic (actual surface) temperature because they are in direct contact. Water emissivity is 0.993, but a thin oil film reduces emissivity to 0.972. Radiant temperature will be different and may be calculated since

$$T_{rad} = \varepsilon^{1/4} T_{kinetic} \qquad \textbf{(eq 6.2)}$$

> **Example 6.1**: For water at a kinetic temperature of 291K, radiant temperature is $T_{rad} = 0.993^{1/4} \times 291 = 290.5K$. For an oil slick, radiant temperature is $T_{rad} = 0.972^{1/4} \times 291 = 288.9K$. The 1.6K radiant temperature difference between oil and water is easily measured by thermal detectors. In thermal images, oil slicks appear cooler than surrounding bright water, figure 6.9. Warm streaks are caused by mousse, which re-radiates absorbed sunlight at thermal wavelengths. Oil slicks and mousse are identified more easily in thermal rather than visible images.

6.11.4 Radar imagery

Radar images also reveal oil spills because oil changes surface radar reflectivity. Oil can reduce radar backscatter as small-scale waves are calmed by even a thin oil layer. Areas of low backscatter are surrounded by stronger backscatter from rough

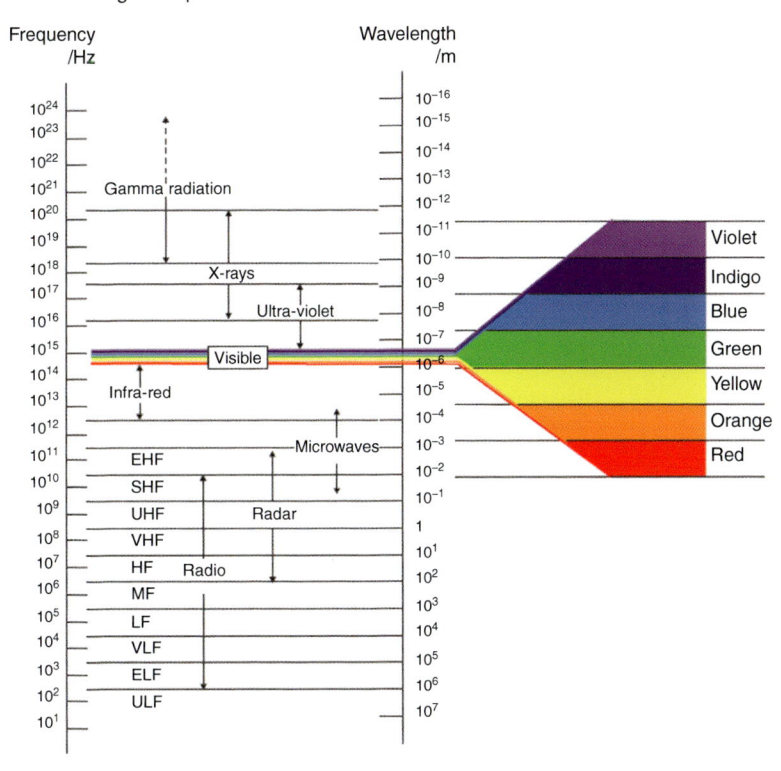

Figure 1.2: *The EM spectrum.*

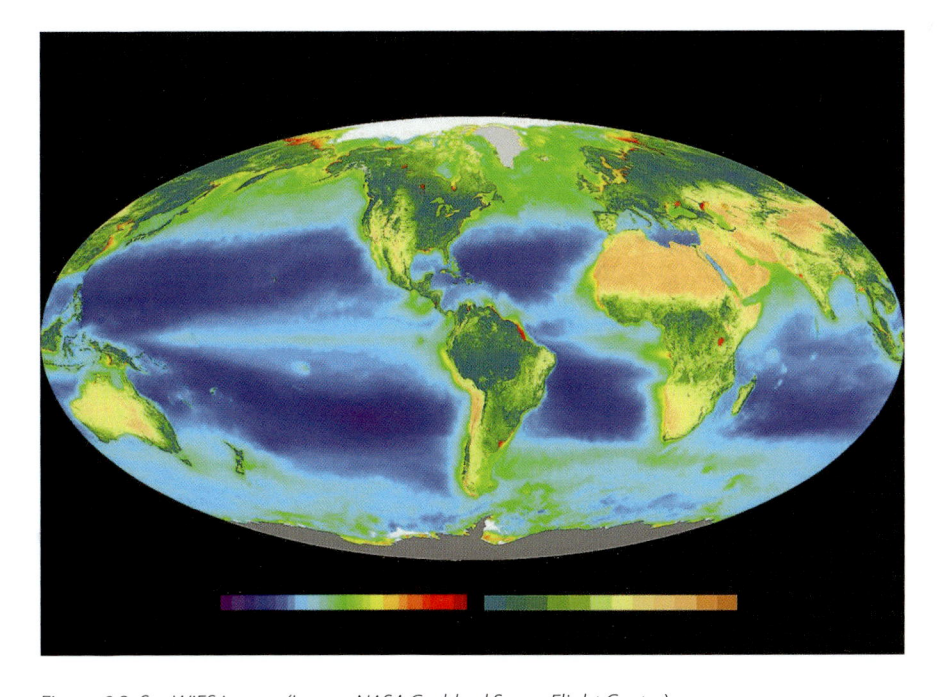

Figure 2.3: *SeaWiFS image. (Image: NASA Goddard Space Flight Center)*

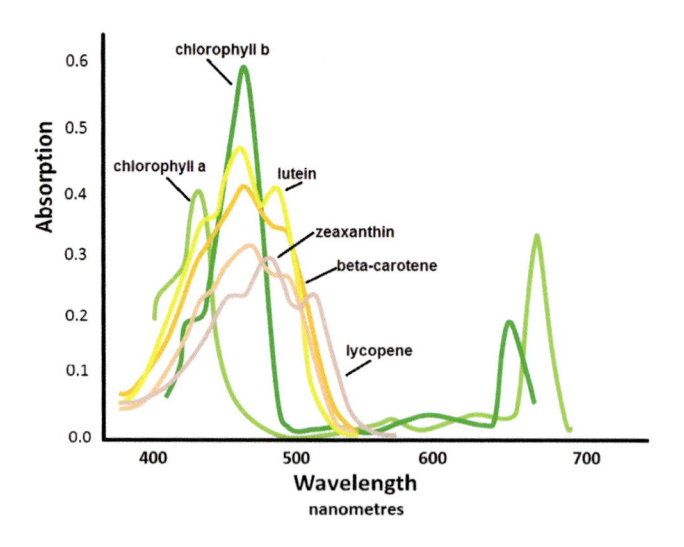

Figure 2.6a: *Scale set for chlorophyll A and B variants (together for illustrative purposes).*

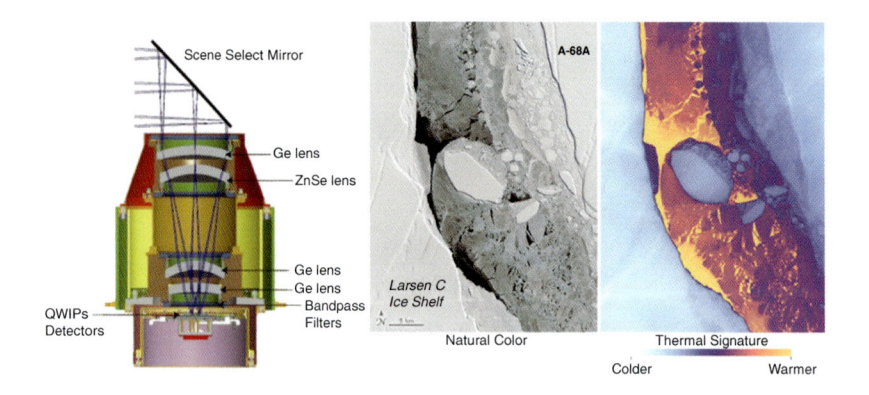

Figure 3.6: *NASA TIRS. The two sensors on Landsat 8, TIRS and the Operational Land Imager (OLI), provide coverage of global land mass at a spatial resolution of 3m (visible, NIR, SWIR) and 100m (thermal). Landsat imagery courtesy of NASA's Goddard Space Flight Center and US Geological Survey.*

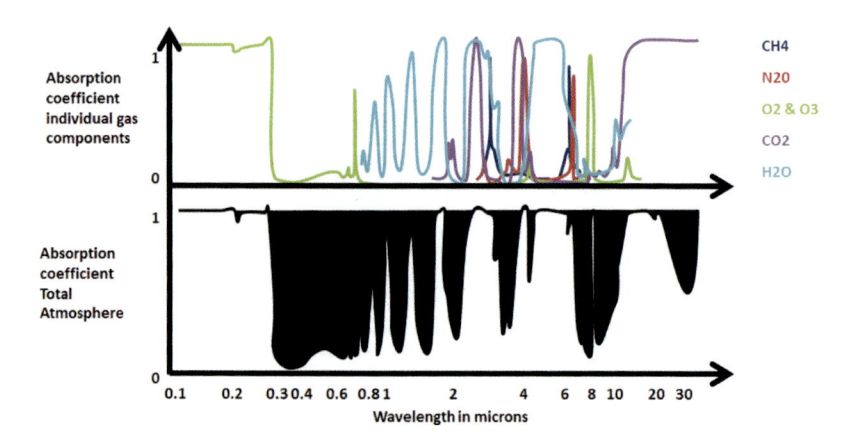

Figure 5.2: *Absorption coefficient individual gas components and total atmospheric content.*

Figure 6.1: *Chernobyl nuclear disaster, thermal imagery. a) left before accident, b) after. Courtesy USGS.*

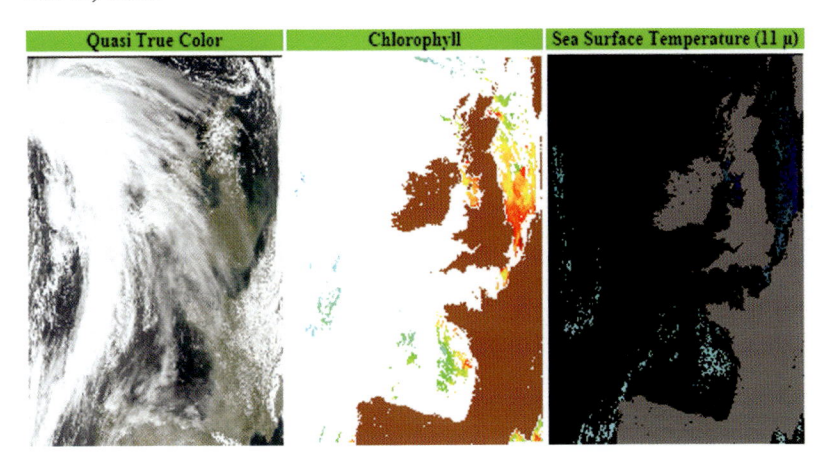

Figure 6.10: *Aqua image: True colour vs chlorophyll vs SST. Courtesy NASA.*

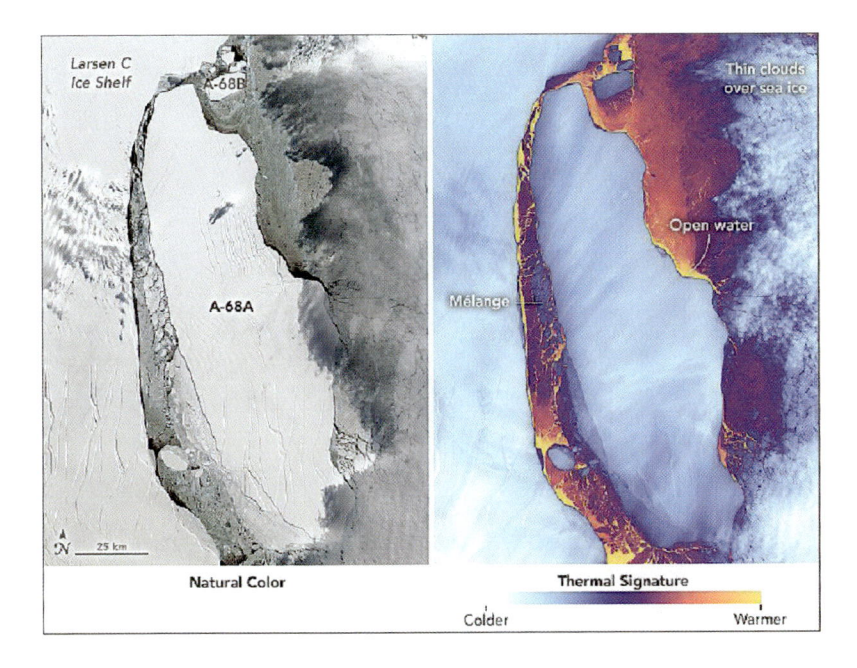

Figure 6.11: *Visible and thermal sea ice signatures [6.14].*

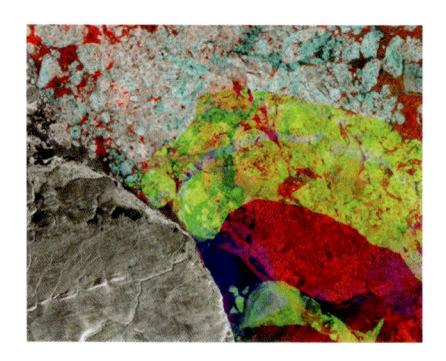

Figure 6.12: *Alert, Canada Copyright contains modified Copernicus Sentinel data (2016–17), processed by ESA, CC BY-SA 3.0 IGO.*

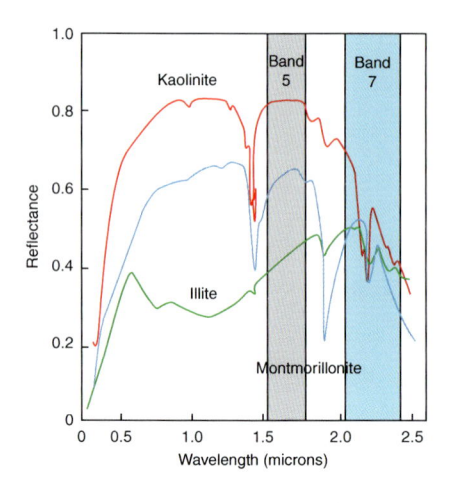

Figure 7.1: *Landsat TM bands 5 and 7 typical reflectance for various clay minerals.*

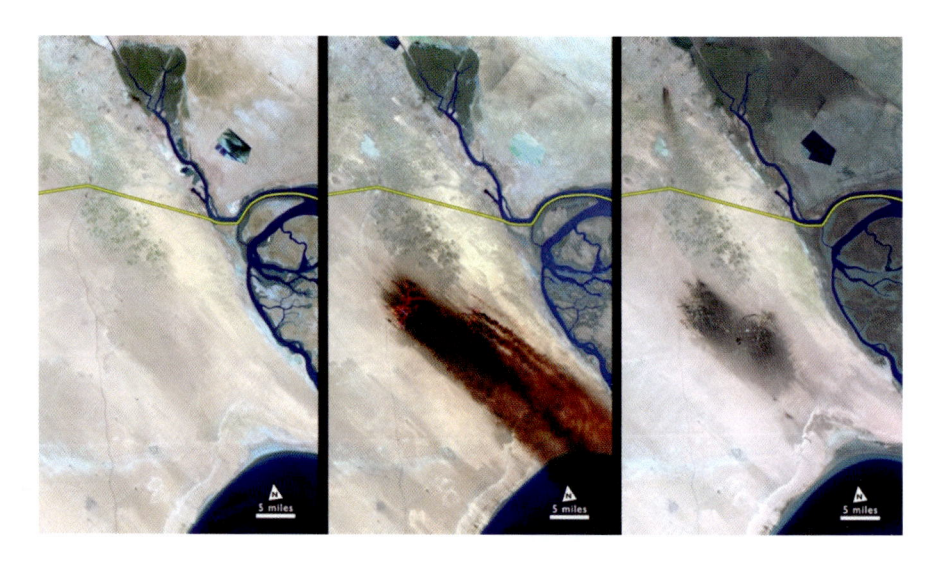

Figure 6.13: *As Iraqi troops withdrew from Kuwait at the end of the first Gulf War, they set fire to over 650 oil wells and damaged many more, just south of the Iraq border (yellow line). These Landsat images show before, during and after the release of 1.5 billion barrels of oil into the environment. Credit: NASA's Goddard Space Flight Center, February 1991.*

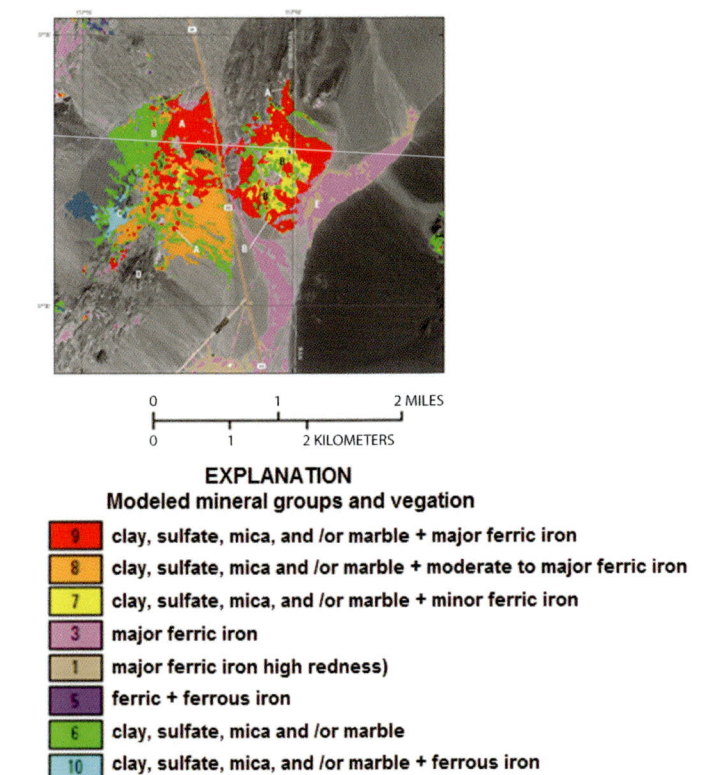

EXPLANATION
Modeled mineral groups and vegation

9	clay, sulfate, mica, and /or marble + major ferric iron
8	clay, sulfate, mica and /or marble + moderate to major ferric iron
7	clay, sulfate, mica, and /or marble + minor ferric iron
3	major ferric iron
1	major ferric iron high redness)
5	ferric + ferrous iron
6	clay, sulfate, mica and /or marble
10	clay, sulfate, mica, and /or marble + ferrous iron
4	ferrous or course-grained ferric iron (may include oxidized basalts, fire ash, some moist soils, and any blue/green rocks)
11	dense, green vegetation

Figure 7.1: *Mineral mapping example [7.6].*

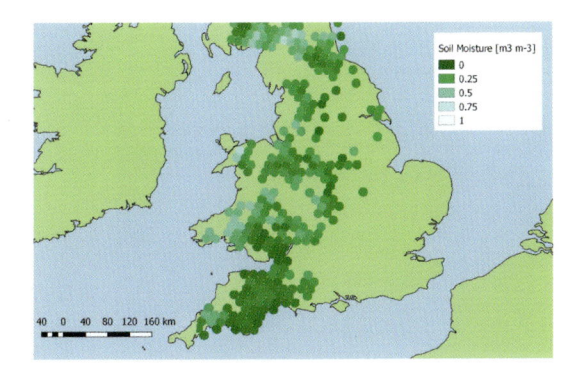

Figure 7.3: *SMOS data courtesy of ESA, with visualisation by Pixalytics Ltd.*

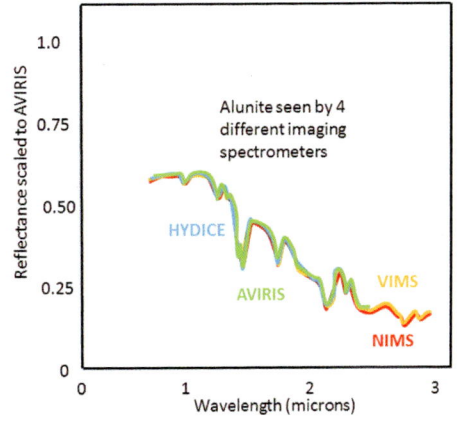

Figure 7.4: *Various sensor responses to the same mineral, referenced to USGS data [7.10].*

April 31, 2000

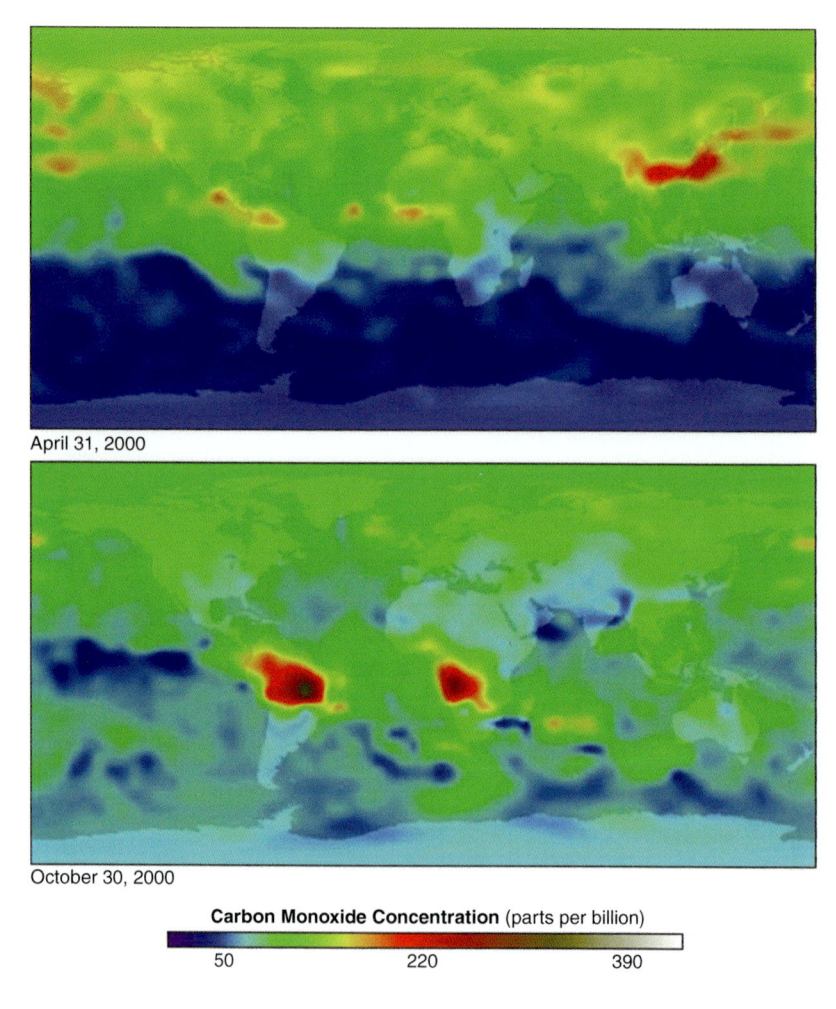

October 30, 2000

Carbon Monoxide Concentration (parts per billion)

50   220   390

Figure 8.4: *False colours represent lower atmosphere carbon monoxide levels from 390 ppB (brown), to 220 ppB (red), to 50 ppB (blue). Author NASA. Source [8.14].*

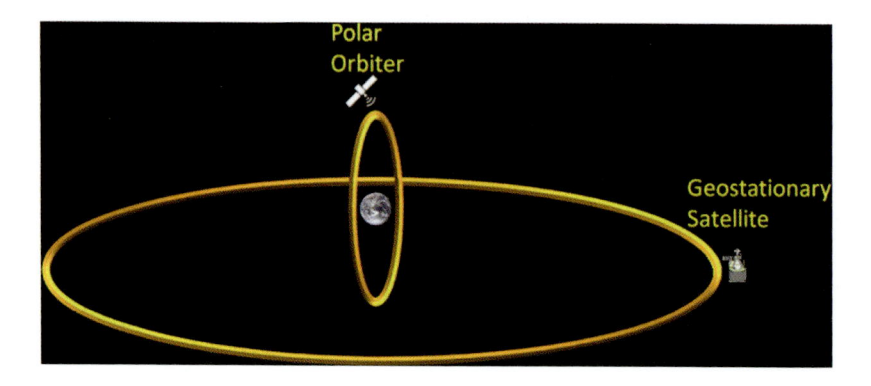

Figure 9.3: *Geostationary satellites and polar orbiters.*

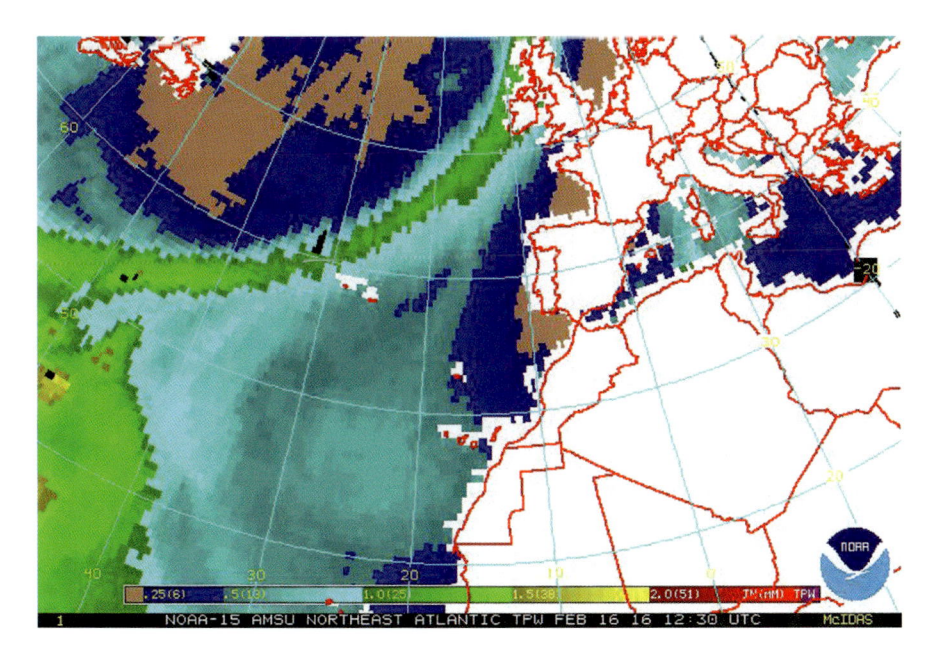

Figure 9.4: *AMSU imagery from NOAA-15 taken 16 February 2016. Credit NOAA.*

Figure 10.1: *Satellite Image of Reykjavik with 'noise' added. Sentinel-2 data courtesy of Copernicus/ ESA, with visualisation by Pixalytics Ltd.*

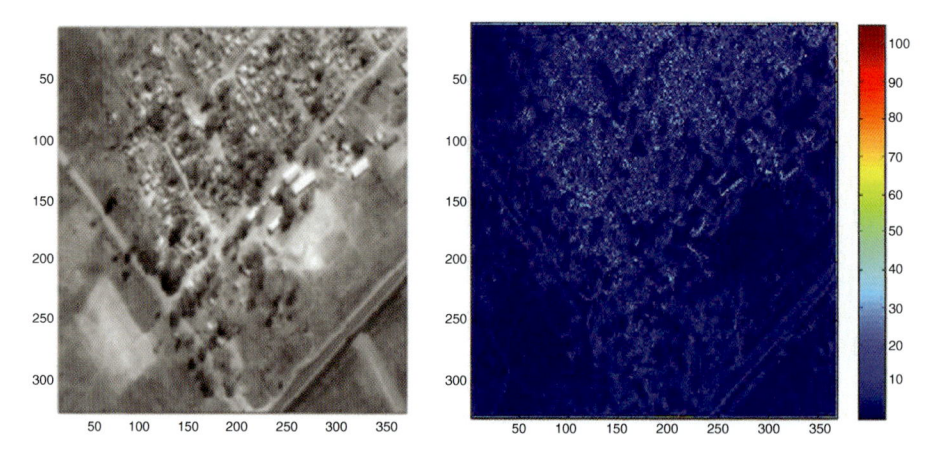

Figure 10.4: *(a) Test image, (b) Absolute display of sliding neighbour operation [10.2].*

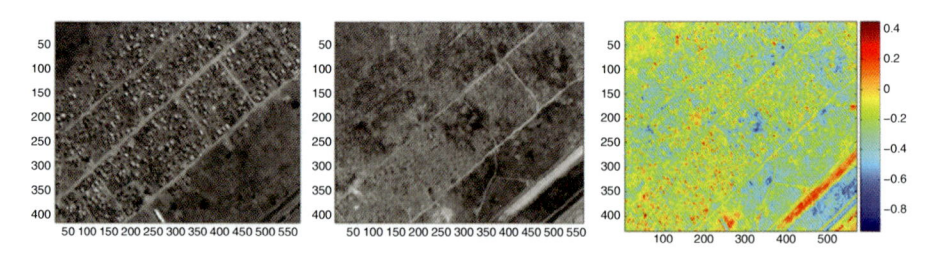

Figure 10.7a: *(left) Before image shows Porta Farm densely inhabited (25 July 2000), while 10.7b (right), after land clearance image, shows the same area (15 September 2006). 10.7c Temporal change between images.*

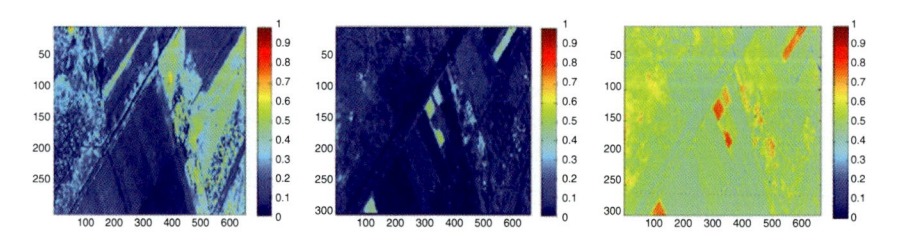

Figure 10.8: a. *Non-radiometrically calibrated NDVI before land clearance, b. non-radiometrically calibrated NDVI after, and c. radiometrically calibrated NDVI for b.*

clean water. Slicks thus show up characteristically dark on SAR images, such as ERS-1 imagery. ERS-1 radar images are sensitive to water roughness differences because of their steep depression angle and especially in VV polarisation. On satellite SIR-A images, slicks are less apparent but enhanced by image processing. Aircraft images are relatively effective at recognising oil slicks. Radar images acquired day or night under all weather conditions are an advantage over other remote systems, including thermal imaging, for monitoring oil spills but are insufficient for large area coverage. However, radar images must be interpreted carefully, as dark streaks may not be oil but simply regions of smooth water!

Oil-covered water is smoother than surrounding water. Many active radar satellites currently provide vital maritime information. The Jason-2 project is a French development programme for operational oceanography. Ocean circulation is studied by measuring sea level height, derived from two elementary data elements: the altimetric distance between the satellite and sea level, deduced from altimetry measurements, and satellite *radial* height in relation to the reference ellipsoid deduced from measurements taken from different positioning systems (see the TOPEX/Poseidon project, which provides similar information [6.15–6.16]).

The longest time spilled oil usually remains in a coastal environment is a decade. The main ecological impact comes at the spillage or within a few months. After the first few months, most of the oil is reduced to tar residues or is chemically detectable in sediments and resident organisms. Impact may be catastrophic in the short term; long-term ecological impact is usually small, as species population recovery in almost every study to date is swift. Inevitably, accidents occur but it is clear every reasonable effort must be made to prevent spills.

6.12 Sea and ice radar interferometry

This is a technique whereby two radar images of the same place are compared. Changes in position of the ground, such as a glacier, are detected with 'repeat pass interferometry'. A second image is acquired soon after the first from a similar position in space. Radar uses coherent bursts of radiation sent out by an instrument, where wavelength and the precise timing of a wave's start are known. When a signal is reflected by Earth's surface, it returns to a receiving system, recording phase and amplitude. Radar images cover vast areas of sea and land and two images of the same place can be acquired at night or under cloud. For example, the active microwave instrument on board ERS-2 had a 100km swath width, acquiring images three days apart. One use is monitoring glaciers, including glaciers extending out to sea. Radar altimeters, such as the ERS radar altimeter, can look at sea-surface topography anomalies, such as El Niño events, viewed over weeks and months, as well as changes at colder latitudes. The Shuttle Radar Topography Mission (SRTM) used C and X-band interferometry to produce the first near-global topographic map of the Earth over its ten-day mission in 2000. The shuttle Endeavour had two radar antennas 60m apart, with over 80 per cent of Earth's surface imaged (60°N to 54°S).

One interferometric radar satellite used for the cryosphere is CryoSat-2, operated to study Earth's polar ice caps. The CryoSat-2 spacecraft was constructed by EADS Astrium and launched by ISC Kosmotras, using a Dnepr carrier rocket, in 2010. On 22 October 2010, CryoSat-2 was declared operational following six months in-orbit testing. Its main instrument is an interferometric radar range finder with twin antennas, which measures the height difference between the upper surface of floating ice and surrounding water, known as 'freeboard'. The general principles of radar altimetry were discussed in Chapter 4. Current satellite missions addressing ocean and sea ice issues include: CryoSat-2, Coriolis, EOS Aqua and Terra, FY-3, GEO-KOMPSat, HY1–2, ICESat, Jason, Meteor M/MP, OceanSat, SAC-D/Aquarius, Sentinel-3, Suomi-NPP and SMOS.

Questions

6.1 Discuss the SeaWiFS system, its operation and applications.

6.2 Why is estimating ocean chlorophyll content important, and how can remote Earth observation satellites contribute? What factors contribute to ocean productivity? Does ocean turbidity affect ocean chlorophyll estimates?

6.3 'For optical passive remote sensing, if the signal is insufficient to collect data in a particular river basin using a 450–550nm band, it is definitely insufficient for a 750–850nm band.' Comment on this statement.

6.4 A scene over water is examined with two thermal IR spectral bands. If T_1 = 318K and A = 0.90, B = 0.15, SST (True) = 323K and C = −15, what is the brightness temperature T_2 recorded passively at 10 microns?

6.5 ICESat has a lidar system for altimetry at 1064nm, and a frequency doubled laser for atmospheric correction.

(i) If the altimeter has a 4.1ms round trip echo time from an ice surface on land when directly overhead, what is the nominal satellite altitude above the ice?

(ii) What is the energy of the frequency doubled photon?

6.6 The emissivity at 37GHz of a coastal land area and nearby ocean are 0.9 and 0.6 respectively. Land and sea are *both* at 303K. What are the possible temperatures observed by a passive sensor at this frequency for land and sea respectively?

6.7 Compare the following space-based radar system capabilities: SRTM, ERS-1 and 2, CryoSat2, TOPEX/Poseidon, RADARSAT and JERS. How does melting Arctic sea ice impact on future commercial and national rivalries?

6.8 Compare visible, thermal and radar image resolving ability for detection of a tanker, small fishing vessel, and a man in the water in terms of detection/recognition/identification, resolving ability, all weather, and day/night capabilities.

6.9 Compare space-based thermal imaging vs terrestrial thermography.

6.10 Discuss the advantages and disadvantages of different terrestrial and satellite sensors that can be used to detect oil spillages on water.

References

[6.1] earthobservatory.nasa.gov/NaturalHazards/view.php?id=79677

[6.2] earthobservatory.nasa.gov/Features/WorldOfChange/aral_sea.php

[6.3] oceancolor.gsfc.nasa.gov/data/czcs/

[6.4] oceancolor.gsfc.nasa.gov/data/seawifs/

[6.5] modis.gsfc.nasa.gov/data/

[6.6] podaac.jpl.nasa.gov

[6.7] 'Use of High Resolution NDVI and Temporal Satellite Imagery in Change Detection for Assessment of Mining, Construction, Environmental and Ethnic Impact', CR Lavers and J Mazower, Proceedings of RSPSOC 2013: Earth Observation for Problem Solving, Glasgow, 4–6 September 2013, ISBN 978162993347.

[6.8] *Recent Developments in Remote Sensing for Human Disaster Management and Mitigation Natural and Man-Made 2013*, CR Lavers (Ed), (Lulu Enterprises, Inc., Raleigh, North Carolina, ISBN 9781291224634).

[6.9] news.nationalgeographic.com/2015/07/150709-iraq-marsh-arabs-middle-east-water-environment-world/

[6.10] www.independent.co.uk/news/world/saddam-drains-the-life-of-the-marsh-arabs-the-arabs-of-southern-iraq-cannot-endure-their-villages-1463823.html

[6.11] *Essential Sensing and Telecommunications for Marine Engineering Applications*, C Lavers (Bloomsbury Publishing, London, 2017, ISBN 1472922182).

[6.12] www.digitalglobe.com

[6.13] directory.eoportal.org/web/eoportal/satellite-missions/s/seasat

[6.14] landsat.visibleearth.nasa.gov/view.php?id=91052

[6.15] sealevel.jpl.nasa.gov/missions/topex/

[6.16] www.ospo.noaa.gov/Products/ocean/ssheight.html

[6.17] www.esa.int/Our_Activities/Observing_the_Earth/SMOS

[6.18] nsidc.org/data/G02135/versions/3

[6.19] nsidc.org/data

[6.20] nsidc.org/data/G02158

[6.21] www.esa.int/Our_Activities/Observing_the_Earth/CryoSat

7
Land Resource Applications

'Landsat represents a public good, Earth-observation infrastructure that allows everyone to study their respective land resources and their change over time.'
Matthew Hansen, 'Landsat Eyes Help Guard the World's Forests', USGS, 3 November 2016 [7.1]

7.1 Land resource applications

In this land applications chapter, we examine specific applications and the land classification process that is required to define land cover for specific uses. We will take the approach in Chapter 6 with a sequence of visible, thermal and microwave where possible, as these three spectral regions provide complementary information. Satellite remote sensing is especially useful for relating observable *patterns* to the processes that cause them. Large-scale processes produce large-scale patterns, while small-scale patterns may go undetected. Regional- and global-scale changes are detected easily by satellites like Landsat, but small-scale variability isn't easily characterised. This is where high-resolution and multi-wavelength visible and NIR sensors prove themselves useful.

7.2 Land cover

In terms of land cover, we will first consider bedrock, material such as leaf mould from which soil is made, soil, then vegetation, and particular applications including archaeological imaging *below* the surface. Earth has a complex surface whose topography (relief) and material composition reflects the bedrock and materials under it as well as changes upon it. Geological satellite and aerial mapping requires field exploration, with the mapping process aided with photographic interpretation. Rock type, fractures, erosion and deposition features provide evidence of the processes that produced them. Through photo interpretation and geological and soil mapping, materials and structures are identified.

7.3 Rocks

The commonest Earth surface rocks are sedimentary rocks, comprising over 75 per cent of land surface. These rocks are formed by consolidated sediment layers settled out of water and converted into rock by lithification, which involves cementation and compaction by overlying deposited sediments. Limestone consists of water-soluble calcium carbonate, and although soils may be well drained, it may be hard to locate groundwater in limestone topography, creating caves, sinkholes and pitted landscapes, such as the Guilin Hills karst towers in Guangxi province, China.

Volcanic *intrusive* igneous rocks form when magma doesn't reach the surface but solidifies in cavities beneath it, pushing rocks apart, melting or dissolving them. Intrusive igneous rocks range from granite (composed of light, coarse-grained quartz, feldspar and mica) to gabbro, a dark, coarse-grained rock of feldspar and ferromagnetic minerals. *Extrusive* igneous rocks form when magma reaches the surface, consisting of lava flows and pyroclastic materials. Flows are generally formed from molten rock exiting volcanic cones or fissures with little explosive activity. Typical igneous landforms, such as Mount St Helens, Vesuvius and Mount Rainier, show light in Ka-band radar imagery with HH, and dark in HV polarisations. Cinder cones and basalt lava flow are rough relative to the wavelength when vertical relief is >1–2cm. Common metamorphic rocks – slate, marble, gneiss and schist – form from sedimentary or igneous rocks by heat and pressure, chemical action or shearing stress. Under high temperature and pressure, rocks undergo *metamorphosis* and new mineral compositions, texture and structure result.

7.4 Geological mapping

The first recorded geological plane mapping helped construct an overlapping view of very many photographs to create the first Libyan 'mosaic' in 1913, with the earliest photos applied to Middle Eastern petrol exploration. Photo interpretation started in the 1920s and since the 1940s, geological evaluation has become widespread, notably in mineral and oil prospecting. Geologic mapping involves identification of landforms, rock type and structure, folds, faults and fractures, and displaying geological units and structures on maps in their correct spatial relationships. Mineral resource exploration is an important geological mapping activity. Most of the globe's surface and near surface mineral deposits have been consumed. Current emphasis is on location of deposits far below Earth's surface or in inaccessible regions. Geophysical methods provide the deep penetration required to locate potential mineral deposits. Information about potential mining sites is provided by correct surface feature interpretation on both satellite and aerial images.

Geological application aerial photos are taken mid-morning to mid-afternoon when the sun has a **high angle** and shadow is minimal. Low sun-angle aerial photos are taken in the early morning and late afternoon when the sun is under 10° above the horizon, with shadow sometimes revealing subtle differences in relief and texture, invisible at higher angles. As important as spectral properties are, it is usually shadow patterns that help geological identification. However, lighting changes rapidly early morning and late afternoon, and lighting angle also varies. Aerial photos are taken by UAVs or planes under snow-free ground conditions because snow obscures ground detail. However, at low sun angles snow may enhance topographic terrain aspects, yielding information not readily extracted from snow-free images. Optical (visible/NIR) images are valuable, but SAR is more useful in frequent cloud-covered regions. Many key significant geological features extend long distances, and are best studied by examining satellite images. An analyst may first examine Landsat images at 1:50,000 scale or high-resolution satellite images.

7.4.1 Geological deposits

Miners have long known that, in mineral areas, mining takes place along broadly linear trends that may run hundreds of miles. In mining areas, fractures and intersections are good prospects as they are avenues for ore-forming solutions in veins or thin sheets. Local fracture patterns are mapped and enlarged on Landsat images, or digitally enhanced in colour to emphasise fractures. Most ore bodies deposit from hot aqueous hydrothermal solutions at a few thousand Kelvin, penetrating host rock, often with unique combinations of metals, water and heat. During ore material formation, these solutions interact chemically with the host rock to alter their mineral composition. Hydrothermally altered host rocks contain distinctive secondary minerals that replace original ones. In regions where bedrock is exposed, multispectral Earth observation can detect altered rocks because their reflectance spectra differ from host rock. Mining extent has been monitored in high resolution in several cases using a red/NIR NDVI approach and different date comparisons [7.2–7.3].

Specific minerals have very different individual and multiple band reflectance characteristics. For example, alunite and hydrothermal clay minerals have distinctive TM band 7 absorption, and higher TM band 5 reflectance. Ratio images emphasise and quantify spectral differences. For example, the TM ratio 5/7 is good for recognising clays. Further detail of the method is found elsewhere [7.4]. From figure 7.1 (see plate section), kaolinite band **5/7** value is 0.57/0.82 = 0.7.

Many indices are applicable to different target mineral detection. For example, ferric iron is detectable using Landsat TM (**3/1**) × ((**3** + **5**)/(**4**)) band ratios and similar to the ASTER (**2**(**1**) × 0.5 × ((**2** + **4**)/**3**) ratio [7.5], showing there is more than one possible approach. However, there is ultimately no completely foolproof way other than empirical testing to see which band comparison methods work best, and human tests to look for particular targets in multiple bands can narrow down to the best solution even in terrestrial imaging [7.6]. Clay-sulphate-mica-marble is found from a straight Landsat 5/7 index ratio. More refined but complicated approaches may be taken (see table 7.1), giving rise to mineral group maps.

7.4.2 Desert sands

Landsat images have shown star dunes in visible/NIR and thermal bands. Longitudinal dunes are imaged in detail with ASTER (with clay between dunes) and aeolian crescent dunes mapped with ASTER and SAR. Loess is an unconsolidated, unstratified, silt-sized windborne deposit with small amounts of fine sand and clay. Loess and loess-like materials cover one-tenth of Earth's surface, comprising silt from desert areas, carried by wind erosion or deposited by meltwater streams from glaciers.

7.4.3 Geological oil exploration

A typical oil exploration programme for searching areas often includes the following: *regional reconnaissance* using Landsat mosaics covering thousands of square kilometres, to locate sediment basins, areas with thick layered sedimentary rocks essential for oil field formation. *Geophysical air surveys* produce magnetic field maps. An aircraft may carry a transmitting coil, while a detector coil passes through an ore body's induced magnetic fields, which generates alternating currents in it. In this way, conductive rock distribution is mapped rapidly over large areas. Sedimentary basins have lower magnetic fields than granite or metamorphic rocks, so gravity surveys show sedimentary rocks have lower specific gravity than granite. Individual satellite image *interpretation* of maps identifies geological structures, such as anticlines and faults, that form oil traps. Likely structures are mapped with stereo satellite or aerial image pairs. *Seismic surveys* using either explosives or powerful acoustic transducers transmit sonar waves into the ground, which reflect

Output Class # and DN Value	Class Colour	Class Name (Materialid attribute)	Material Index					
			3/1	3/1 × (3 + 5)/4	(2 + 5)/(3 + 4)	5/7 – 4/3	4/3
1	light brown	minor ferric iron (high redness)	X				
3	magenta	major ferric iron	AND	AND			
4	dark cyan	ferrous iron or oxidized basalts (may include any blue-green rocks, fire ash, and some moist soils)			X		
5	purple	ferric + ferrous iron	AND/NOT	AND	AND		
6	green	clay, sulfate, mica, and (or) marble				X	
10	cyan	clay, sulfate, mica, and (or) marble + ferrous iron			AND	AND	
7	yellow	clay, sulfate, mica, and (or) marble* minor ferric iron	AND	AND		AND	
8	orange	clay, sulfate, mica, and (or) marble* moderate to major ferric iron		AND	AND/NOT	AND	
9	red	clay, sulfate, mica, and (or) marble + major ferric iron	AND	AND	AND/NOT	AND	
11	dark green	dense, green vegetation					X

Table 7.1: *Band classification key to figure 7.1.*

from underground anticlines. Echoes are processed to provide subsurface detailed seismic maps. Finally, wells are *drilled*.

7.5 Soil mapping and evaluation

Many soil aspects were covered in Chapters 1 and 2. Detailed soil surveys are a key source of resource information, and are used extensively in planning land use. Understanding soil suitability for various land activities is essential to prevent environmental deterioration associated with land misuse. Soil scientists may need to visit areas, identify and mark soil boundaries and produce detailed surveys. This process involves field soil examination, identification and classification, depending heavily on experience and training to evaluate the relationship between soils and vegetation, parent rock material, landform and landscape. Soil survey maps have been prepared for the US Department of Agriculture since 1900, providing technical assistance to farmers and ranchers for crops and grazing animals. Air photo interpretation has been used since the 1930s for soil mapping, with panchromatic aerial photos. Since 1957, soil surveys have shown information about suitability of mapped soil for various uses. These surveys contain information for estimating crop yields and woodland productivity and assessing habitat conditions, recreational land use and suitability for development. Sunlight reflection from bare soil depends on many factors, including moisture content, soil texture, roughness, iron oxides and organic matter content. One area of bare soil may have different photo 'tones' on different days, depending especially on moisture content. As vegetation cover increases during the growing season, reflectance should be more due to vegetation than soil (see Chapter 2). African savannah has just about every soil type, but can dry out, so fires are common. African soils have been mapped extensively by various researchers [7.7] and maps are available online [7.8].

Soil characteristics are important– but to whom? Soil photo interpretation helps farmers, but can also indicate underlying mineral deposits or oil reserves to prospectors, while terrain characteristics include bedrock type, landform, soil texture, drainage, flood risk and depth of unconsolidated material over bedrock. Soil is composed of two layers, a strongly weathered surface **topsoil** typically 0–0.6m thick, mostly fine particle organic matter. Next is the **subsoil**, 0–2.5m thick, containing some organic matter with accumulations of fine-textured particles washed down from the topsoil. Below is the underlying geological rock from which topsoil and subsoil develop, the so-called **parent material**. There are three primary soil sources: *residual* soils formed from bedrock redistributed by natural weathering, *transported* soils redistributed from parent material by wind, water or glacial ice, and *organic* soil peat from decomposed plant materials in wet environments, such as bog areas with a high water table. Soil is a mix of solid particles, water and air. Particles are given names like gravel, silt and clay, based on size: gravel 2.0–76.2mm, sand 0.05–2.0mm, silt 0.002–0.05mm, and clay below 0.002mm. Materials >50 per cent silt and clay are fine-textured, while materials >50 per cent sand and gravel are coarse-textured. Texture names are given to specific mixes of sand, silt and clay.

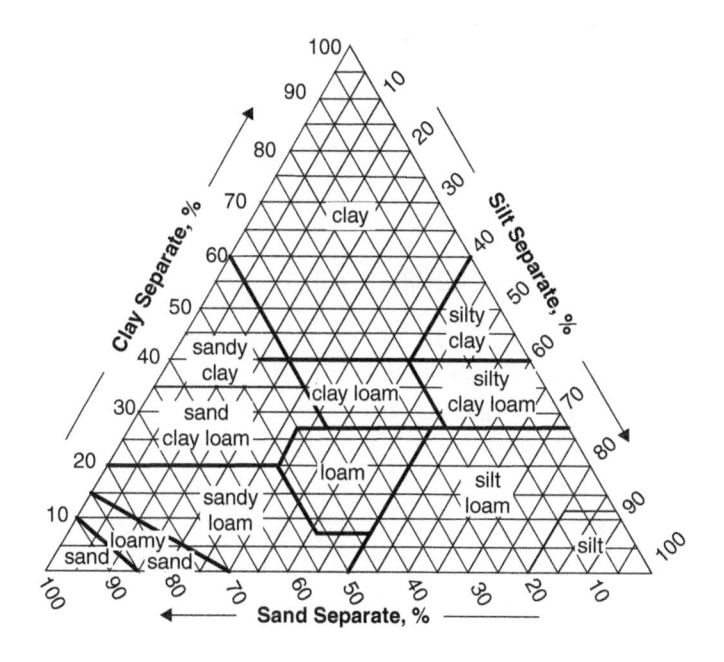

Figure 7.2: *Soil triangle [7.9].*

7.6 Soil salinity

Saline soil is recognised as a limiting factor to vegetation growth. Increased irrigation draws salt-bearing minerals into groundwater. Evaporation of salt-bearing irrigation water also increases soil salinity. The 'drawdown' of coastal aquifers, especially in Middle Eastern and Mediterranean climates, allows salt water to intrude into groundwater. The NASA SMOS satellite provides data on both sea and land, and benefits soil scientists and farmers, see figure 7.3 (see plate section). Units are soil volume in m^3 per m^{-3}. The SMOS mission has three arms with 69 detectors, operating as a 2D interferometric radiometer. It has a spatial resolution of 35–50km, depending on the swath location within the instrument's FOV. Being hexagonal in shape, computer code approximates circles that are sub-sampled at 1km resolution.

Landsat TM colour composites with bands 2–3 and 7 (visible and NIR) are useful to monitor sea-salt covered soils. *Thermal* data from platforms like the Daedalus Airborne DS-1260 MSS thermal infrared (TIR) provides imagery whose brightness depends on soil moisture content (less change = more moisture, more change = more salt). *Radar* returns depend on dielectric constant, and salt has a high dielectric constant. The higher the moisture content, the higher the observed SAR radar backscatter. Dry soil (desert) has low dielectric, forest and moist soil has moderate dielectric, while saturated grassland and soils are high. Tundra (frozen/unfrozen and permafrost) has variable dielectric. Many minerals (section 7.4.1) including iron oxide content have been mapped with satellites like AVIRIS, along with organic matter content. It is emphasised that different sensors and satellite-based spectrometers do *not* see the same material in exactly the same way [7.10], see figure 7.4 (see plate section).

7.7 Land use/cover mapping classification

Land use and cover are vital for planning and management activities concerned with Earth's surface. Many organisations require land cover and want to know where use is changing, to what extent, and the timescales involved. Users want to understand the causes and consequences of change and whether they can project potential impact, for example to houses on floodplains. Today we are also concerned with climate change impact. Satellites provide images on a global scale: with MODIS/AVHRR, local/regional Landsat TM/ETM + and SAR. Much land cover and vegetation datum is available freely. MODIS vegetation products are available at 500m resolution [7.11]. Panchromatic medium-scale aerial photo mapping for land use has been accepted practice since the 1940s. Today, small-scale aerial photographs and satellite images provide land use/cover mapping over large areas. Land cover relates to the features present on a surface. Urban buildings, lakes and trees are examples of land cover. The term 'land use' relates to human activity associated with the land. The US Geological Survey (USGS) devised a systematic land use/cover classification system for use with acquired remote sensor data. The USGS Land Cover Classification System, Level I and II, used with Earth observation data, is viewed in table 7.2. USGS Land Cover products are available [7.12].

Urban or built-up land is covered by man-made structures, such as transportation, power and communication facilities, shopping centres, industry, commercial and man-made institutions. The urban category takes precedence over all others when criteria for more than one category are met. For example, residential areas with

Level I	Level II	
100 Urban or built-up land	110 Residential 120 Commercial and services 130 Industrial 140 Transportation	150 Communications and utilities 160 Institutional 170 Recreational
200 Agricultural land	210 Cropland and pasture	230 Confined feeding operations 240 Other agricultural land
300 Range	310 Herbaceous range 320 Shrub and brush range	330 Mixed range
400 Forest	410 Deciduous forest 420 Evergreen forest	430 Mixed forest 440 Clear cut areas 450 Burned areas
500 Water	510 Streams and canals 520 Lakes and ponds	530 Reservoirs 540 Bays and estuaries 550 Open marine waters
600 Wetland	610 Forested wetland	620 Non-forested wetland 630 Non-vegetated wetland
700 Barren land	710 Dry lake beds 720 Beaches 730 Sand and gravel areas other than beaches 740 Exposed rock	750 Strip mines, quarries and gravel pits 760 Transitional areas 770 Mixed barren land
800 Tundra	810 Shrub and brush tundra 820 Herbaceous tundra 830 Bare ground tundra	840 Wet tundra 850 Mixed tundra
900 Perennial snow or ice	910 Perennial snowfields	920 Glaciers

Table 7.2: *USGS Land Cover Classification.*

tree cover sufficient to meet forest criteria are placed in the urban land category. A Level I land-use classification map, interpreted from Sentinel-2 imagery, of Grimspound, Dartmoor, UK is shown in figure 7.5 (left), with an interpretive Level I land classification map (right).

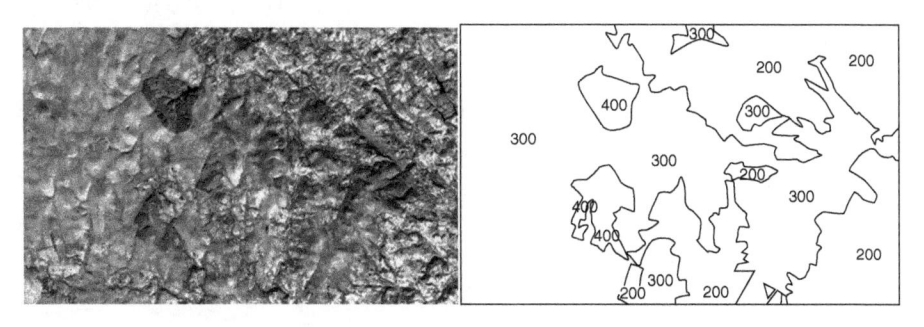

Figure 7.5: *Dartmoor (left) Sentinel-2 data courtesy of Copernicus/ESA, with visualisation by Pixalytics Ltd, and (right) land classification.*

7.8 Vegetation cover

The physical basis for vegetation spectral reflectance characteristics was covered in Chapter 2. It is worth commenting that differences in information available from various sensors combined with the unique vegetation response (spectral resolution) play a big role in identification with hyperspectral, multispectral and radar sensors. Vegetation patterns and processes can appear different depending on spatial scales and so resolution is important. Temporal resolution is critical to understand annual vegetation changes through its growing cycle.

Visible Satellites such as AVIRIS readily distinguish crops separately by considering spectral signature (0.4–2.5 microns). Reflectance evaluates plant water and transpiration so relative water content is obtained. Surface effects are observed on bidirectional reflectance differences between farmland and forest with band composites and viewing angle. Farmland provides homogeneous land surface cover, while snowy forest is heterogeneous. Vegetation indices were introduced in Chapter 2, alongside simple formulae to estimate vegetation. Most indices take the red versus NIR reflectance differences of green vegetation, which reduces atmospheric illumination variation effects. The Leaf Area Index is useful on a global scale, for example with applications to the southern hemisphere such as southern Africa, providing NDVI maps for land clearance [7.13]. NDVI is proven in differentiating one desert area from another, looking at NDVI differences for the Negev and Sinai deserts, with daily rainfall [7.14].

Thermal This band is useful, as thermal emission and plant water stress are related. Emission measurements help derive crop surface temperature. As water transpires from plants, leaves cool due to evaporation. If plants are stressed, leaf temperatures increase. Aquatic plants are also studied, as they still absorb light and NIR while emitting thermal radiation (warmer than surrounding water) and are detected in thermal imagery.

Microwave The penetration depth through a vegetation canopy varies with wavelength. Direct backscatter is achieved from vegetation with 3cm waves (X-band), typically leaf-sized. Direct returns are possible from soil surfaces below the canopy with >24cm (L-band) waves, which see through vegetation, ideal for military surveillance. Returns from plants and soil together are achieved through multiply scattered C-band (6cm) waves.

7.8.1 Agricultural crop yields

Agricultural Earth observation applications are varied, as the physical, biological and technological problems facing modern agriculture are wide ranging and related to global issues such as population, energy, environmental quality, climate and weather. These factors are influenced by culture and economic, political and social systems. Three basic criteria for agricultural yield calculations rely on remotely sensed data: (i) *crop type classification*, based on the assumption that specific crop types are identified by spectral response patterns and image texture; (ii) *crop condition assessment*, where poor conditions – including crop disease, insect damage and plant stress – are assessed and should be treated; and (iii) the actual *yield estimation*.

7.9 Forest applications

Vegetation land use is defined more precisely at Level II than Level I – for example, classifying woodland Deciduous 410 or Evergreen 420 rather than Forest 400 at Level I. Forestry is concerned with woodland, wildlife and recreational management. Species identification is more complex than agricultural crop identification, as forests are often composed of a tree species mixture as opposed to monoculture agricultural land. Tree species are identified on photos by first eliminating species whose presence is unlikely due to geography, location or climate. The second stage is to establish which species don't occur in the area, based on knowledge of common species and their requirements. Photo characteristics – shape, size, pattern, shadow, tone and texture – are used (Chapter 10). Individual species have characteristic crown shapes and sizes. Some species have rounded crowns, while others, such as conifers, are more pointed or cone-shaped. When trees are isolated, shadows may provide profiles useful for identification. The extent to which tree species are recognised on photos is determined largely by scale and photographic quality. Particular tree characteristics, such as crown shape, colours and branching patterns, are essential for successful identification on large-scale photos. Correct interpretation becomes worse as scale is decreased. Historically, the format widely used for tree identification is panchromatic at 1:20,000 scale. Black-and-white IR prints are used to separate coniferous trees from deciduous.

Seasonal change helps discriminate species otherwise indistinguishable. The obvious example is deciduous and coniferous separation, when deciduous foliage falls in the autumn and is reversed during spring. Interpretation helps determine harvested timber volume, which depends on tree or **stand** height, crown diameter and stock density. The height of individual trees is determined by measuring relief displacement, or *image parallax*. Tree volume is determined as a function of species, crown diameter and height. Again, use of the 'red edge' and NDVI using red/NIR bands, including NIR photography, can prove useful.

7.10 Archaeology

Archaeology and ancient riverbed detection is important to the scientific study of prehistory, examining remains discovered through excavation. The first archaeological investigations dealt with monuments of early societies. Surface features include visible ruins, mounds, rock piles and other surface markings. Examples of visible ruins are rock structures, such as Stonehenge, and European castles, although recent ancient discoveries such as Gobekli Tepe, Turkey were buried and *not* visible from either air or satellite. Examples of visible surface markings include Indian pictographs and the ancient Nasca Lines, Peru (figure 7.6), estimated to be some 1,500 years old and covering 500km^2. Many geometric shapes were found, as well as narrow straight lines extending 8km. They were made by clearing millions of rocks to expose lighter-toned ground. Cleared rocks were piled round the lines' outer boundaries. These markings were first noticed from the air during the 1920s and mapped extensively by Paul Kosok, a New York historian, and Maria Reiche, a German-born Peruvian archaeologist.

Figure 7.6: *Nasca lines, © Landsat*.

In the century since archaeological photographs were first taken, photography and now satellite imagery has become a reliable method for surveying archaeological sites. Traditionally, photos were obtained from light, low-flying aircraft taking vertical or oblique photographs. With the advent of high-resolution satellite imagery and close-range UAVs, archaeologists have found other sources to look for traces of ancient inhabited sites. Features may range in size from earthworks kilometres long to soil and crop marks under a metre wide. Earthworks are features seen in relief, appearing on aerial photos as highlights and shadows with vegetation cover differences. Soil marks can reveal buried ditch banks or foundations through soil colour changes, particularly when buried features are brought to the surface by ploughing. Buried walls and ditches are detected as crop marks where buried features decrease crop growth as a result of differences in moisture and nutrient availability.

Visible The first of a new generation of commercial high-resolution remote sensing satellites useful to archaeology became operational with the launch of IKONOS in 1999. Providing 1m panchromatic and 4m multispectral sensors, including NIR, and tasked at specific times, IKONOS and satellites like it have great archaeological potential. Landsat TM also has an NIR band to detect archaeological features through subtle vegetation cover differences, but IKONOS has superior resolution.

Thermal Thermal inertia mapping has application to archaeological surveys. If one material is buried within another and the materials are different, heat flow may generate surface temperature anomalies, yielding information on buried objects. One modern thermal application of particular importance to the military and civilians worldwide is detection of buried mines.

Active radar Microwave radiation penetrates vegetation or soil if conditions are right, using long wavelengths, and where vegetation or soil is dry, showing information about features hidden from the human eye. Radar images are harder to interpret than photographs, due to two factors: layover and speckle, mentioned previously (Chapter 4). Subsurface satellite radar imaging is possible if topographic cover is radar smooth and the penetrated material fine-grained – that is, under a few metres thick and very dry. Where cover is less than *skin depth*, echoes increase because of wave refraction and reduced oblique incidence backscatter [7.15], evaluating radiation penetration into Earth's surface (0.1–1.5GHz) for remote sensing applications. The first SeaSat L-band radar showed igneous dykes beneath the Mojave Desert buried under 1–2m of alluvial cover. In 1984, due to the hyper-arid conditions, satellite radar on the space shuttle SIR-A detected Saharan desert dry riverbeds, where sand skin depth can be > 5m.

The first NASA imaging radar SIR-A and SIR-B images acquired over Egypt and Sudan demonstrated space-based 24cm L-band SAR ability to return geological information from 1–2m depth from loose sand cover and semi-consolidated alluvia. The area of south-west Egypt and north-east Sudan is uninhabited and, until 30 years ago, poorly known, and is considered unusual because it lacks surface drainage features. Scattered, extensively wind-eroded outcrops lacked fluvial patterns either on the ground or with conventional visible/NIR images such as Landsat and SPOT. The radar sensor's view of buried river valleys and surrounding fluvial topography arose because surface sand in this part of the Sahara is underlain by secondary calcium carbonate deposited in valley sediments during periods wetter than at present. Natural terrain subsurface imaging with long wavelength SAR from space or air-based platforms – SeaSat, SIR-A/B/C, ERS-1–2, JERS-1, RADARSAT, TRMM and SRTM – is demonstrated where topographic surfaces are radar smooth, appearing dark on radar images, and where shallow subsurfaces contain dielectric interfaces and variations, which appear bright. Similarly, radar penetration from space is applicable over 10 per cent of Earth's surface – for example, the deserts of China and Saudi Arabia and the Mojave Desert. Depth penetration is proportional to wavelength and inversely proportional to the dielectric, and best in clay-free minerals (these hold absorbing water). Microwave remote sensing provides a unique ability to monitor land processes, generally through development of airborne and space-based microwave sensors.

7.11 Land glaciers *visible*

ASTER images of receding glaciers and other glacial landforms have been monitored in many inaccessible regions such as Bhutan, where lakes form at glacier termini. Our own work with West Papuan glaciers using IKONOS and Pleiades data shows that visible and NIR satellite bands provide high-resolution decadal imagery [7.2]. Landsat TM also provided evidence of massive glacial erosion of bedrock, especially in Canada.

Thermal The value of thermal systems to monitor sea ice has been discussed in Chapter 6 (snow and sea ice thermal signatures) and is also important for land ice applications.

Microwave Typical L-band HH interferometric SAR has looked at glaciers such as the Lambert Glacier, Antarctica. NASA used this data to provide ice velocity values. Shuttle SRTM C-band and X-band interferometry also produced the first near-global topographic Earth radar mapping.

7.12 Urban and regional planning applications

Urban planners use data to formulate government policies and programmes, for social, economic and cultural, environmental and natural resource planning. There is need for planning agencies to have timely, accurate and affordable data. Population estimates are obtained through aerial photos from dwelling unit number of each housing type and then multiplying the number of units by average family size per dwelling [7.13]. Many environmental factors affect housing quality, and are interpreted from aerial photos using individual and multiple spectral comparisons of visible and NIR bands, while others cannot be directly interpreted. Today, nearly 50 per cent of Earth's population lives in cities, rapidly expanding at their peripheries. This has a huge impact on land cover, societal structure, population distribution, land use characteristics and so on, as well as straining critical infrastructure, which may lag expansion. Useful satellite sensors for urban classification include aerial photography and satellite imagery: Landsat TM and ETM + , IKONOS, QuickBird, lidar, hyperspectral (CASI, AVIRIS), MOPITT and TOMS (atmospheric pollution) and ASTER. In urban environments, there is a complex mix of vegetation, water, concrete and so on. High spatial resolution data is needed, while temporal and spectral resolution isn't required for most applications as man-made surfaces don't change greatly. Quantities used to measure urban extent include housing density, structure type, urban vegetation cover, air quality and various change detection methods. Delineation of urban areas is difficult because urban is complex and diverse, and boundaries between urban and suburban aren't always clear. There is also a lack of consensus on a consistent definition of what urban actually is. Landsat TM multispectral image data shows differences between well-developed city centres, indistinct ones, or between urban and natural environments. MSS and ASTER looked at these problems. Urban classification is difficult due to mixed pixels, but the USGS system helps address such issues.

7.12.1 Monitoring urban growth visible

Traditional census data lacks spatial details and is updated every decade. Earth observation data use for change detection requires high spatial resolution and careful image co-registration. Remote sensing can monitor urban growth in developing countries more frequently.

City lights The visible and NIR bands of the Operational Linescan System, on board the DMSP satellite, generated a global inventory of human settlements recorded into 64 levels at high (0.55km) and low (2.7km) resolutions. With improving social standards, the more developed and wealthy a region, the greater the correlation with street and dwelling lights. Spatial resolution of 1km is achieved, with relationships between city lights and socio-economic variables such as population density, economic activity, electric power consumption and so on.

Thermal Urbanisation leads to changes to a city's surface energy balance, creating significant urban heat island effects, as well as changes to air quality. For example, thermal images of a city in day and night can be very different and informative. Demographic/socio-economic patterns are important, as you need to integrate physical and socio-economic variables. Pozzi and Small (2002) showed relationships between population density from US census data, and vegetation cover from Landsat TM, finding a linear inverse correlation between population and vegetation fraction [7.16].

Radar urban mapping Shuttle Radar Topography Mission (SRTM) data collected in February 2000 with 30m spatial resolution observed high radar backscatter from urban areas, deriving its extent, urban/suburban vegetation height and distribution, building height and volume, which, combined with other data, helps quantify economic development, transportation infrastructure and so on.

7.12.2 Terrestrial building heat surveys

Aerial and UAV thermography studies have evaluated heat loss from buildings worldwide. The method reveals differences on roofs. Images help assess damaged insulation and roofing materials. Thermal scanning estimates heat loss and is best conducted on cold winter nights, several hours after sunset, or cold overcast winter days, to minimise solar heating effects. Roofs shouldn't be snow-covered (insulating) or wet (thermal washout). Wide-scale thermal IR images were acquired during several unmanned satellite programmes: Landsat TM, AVHRR and HCMM mission, besides other environmental and meteorological satellites, providing city-scale thermography.

7.13 Land surveillance

Land surveillance is required by military, paramilitary (police and law enforcement), as well as search and rescue and civilian users from unmanned and manned platforms. Surveillance, reconnaissance and targeting are critical for armed forces with real-time situational awareness on land as well as at sea (Chapter 6). Flexibility in surveillance range and sensor selection is essential to mission success. The ability to instantly process and exchange data from the field is a vital component to achieve terrain dominance, or a successful emergency rescue for a civilian. Systems must operate across a range of scenarios, combining day and night sensors, radar, laser range-finders and unmanned airborne systems with digital navigation maps, and ground-based sensors (electro-optics and seismic), to provide comprehensive surveillance cover. Land mission systems include automatic intruder detection, IR night vision surveillance, homeland, border and perimeter security, bird strike protection, weapon sites, nuclear plants, petrochemical installations, ports and airports. Many systems provide short-range monitoring from carefully sited towers or buildings. For greater range, tethered aerostats (airship blimps) provide detection and tracking options over challenging terrain, with cutting-edge optical, thermal and radar sensors. Systems must see in total darkness, through obscurants, maximise detection, and operate where lighting is poor or best avoided. There are many applications where thermal imagery can benefit organisations – for example, thermal cameras have been used to secure the perimeter of BASF, Germany and Norwegian electrical substations, enhance UK solar farm surveillance or provide home protection. There are many issues surrounding privacy protection with increased surveillance camera use in public areas, and hard to spot UAVs, but they provide opportunities for current and future requirements.

Longer range land surveillance from space is a critical requirement across military and civilian security sectors with applications such as threat detection, terrain mapping and disaster prevention. A key technology for land surveillance, SAR provides high-resolution radar images in all weather conditions. Recent Interferometric SAR (InSAR) and Differential Interferometric SAR (D-InSAR) have become powerful tools adding high-resolution elevation and change detection, providing 1m ground resolution, while airborne counterparts obtain 10cm resolution. D-InSAR systems can produce cm-scale vertical image products. Disaster land surveillance applications – land subsidence monitoring, landslide hazard prediction and tactical target tracking – may benefit from improved resolution. However, the key resolution limitation of an imaging system is wavelength. One system offering potential for improved resolution is Synthetic Aperture Ladar (SAL), a system operated in the IR, much smaller than radar wavelengths. Future systems may operate at 1.55 microns and integrate on-board optronic processors, providing transmission of high-resolution images to ground end users. More advanced surveillance options are becoming available, including fusion of radar and electro-optical signals to reduce false alarms.

7.14 Disaster monitoring

All disaster types have been widely monitored on land, as this is where natural and man-made disasters have the greatest impact on human populations. Disasters impacting land include disease, hurricanes, tornados, earthquakes, floods, war, terrorism and so on. Imagery of such events is widely available and will not be repeated here.

Geospatial mapping of disease or epidemiology includes cholera, attached to zooplankton and phytoplankton. Plankton plumes from the Ganges are regularly monitored for this. Hantavirus, carried by mice, correlates with precipitation changes such as El Niño. Townsend found that Ebola outbreaks correspond to land use and seasonal climate pattern changes [7.17].

Man-made fire mapping can estimate burned biomass (tropospheric ozone, CO_2, CO etc) and is covered in Chapter 8). Thermal satellite bands provide active fire detection and characterisation but at poor resolution. MODIS burned area mapping is available with 1m pixel resolution versus ASTER's 90m resolution, and is available freely [7.18]. ATSR also helped provide a world fire atlas from night-time data.

Desertification Since 1984 AVHRR has provided Saharan mapping, which moved south 240km in 1980 to 1984 but has fluctuated greatly since. US high plains sand dunes have recently 'reactivated', moving perhaps in response to climate change. Landsat TM/ETM + has extensively looked at sand dune migration.

Forest clearance is monitored with multispectral as well as SAR imagery, with Amazonian rainforest deforestation mapped extensively using JERS-1, which penetrates cloud cover, providing 30m resolution. Landsat TM provides multispectral vegetation NDVI mapping and albedo changes. SAR backscatter can monitor deforestation, especially in regions with persistent cloud cover.

Earthquake radar interferometry The ground we walk on may appear relatively stable, but in active zones it is subject to sudden and violent movements due to tectonic forces generated by Earth's plates. Traditional methods of measuring ground movements rely on taking equipment into potentially dangerous areas for making observations in situ, yet it is well known that measuring small changes in ground positions is possible from space. If we study ground motion detail from an earthquake or ground swelling beneath active volcanoes, patterns may help predict earthquakes and volcanic eruptions. Displacements are measured from space with interferometry. ERS-1 and ERS-2's first tandem interferometric black-and-white pictures provided dark low coherence and bright high coherence for such changes. Images reveal small interferometric ring movements of ground due to seismic activity associated with Mount Vesuvius, or larger volcano movements such as the Yellowstone supervolcano.

7.14.1 Humanitarian aid for natural and man-made disasters

UNOSAT and AAAS have extensively mapped disaster areas to support relief workers and NGOs engaged in relief support, with an extensive archive. UNOSAT delivers imagery analysis and satellite solutions to relief and development organisations within and outside the UN to help in critical areas such as humanitarian relief, human security, strategic territorial and development planning [7.19].

Man-made disasters High-resolution satellite imagery is used for human rights-related monitoring. Imagery is useful for assessing the extent of violent conflict, forced displacement and other human rights concerns in remote, inaccessible or otherwise oppressive state-controlled parts of the globe. As high-resolution satellite imaging capabilities have developed, so too has the power to analyse the conflict impact on infrastructure and other features identifiable from imagery. In addition, the decreasing cost of geospatial technologies and increasing availability of geospatial data have made high-resolution imagery analysis a reliable tool for human rights organisations [7.20]. Satellite imagery can be used to monitor controversial engineering developments, such as dam building [7.2], showing construction of camps for site workers, and mining infrastructure [7.1]. It also provides highly detailed imagery of individual vehicles at these sites. High-resolution land clearances were monitored in Zimbabwe, providing reliable estimates of displaced people and NDVI products [7.13].

Questions

7.1 State the difference between land cover and land use, and why land classification is important for planning use.

7.2 Investigate current trends in the use and consumption of mineral products, and consider the likely consequences of continued exponential growth in exploitation of non-renewable mineral deposits.

7.3 Explain the importance of the different stages of a remote sensing survey to geological mapping natural gas and oil exploration.

7.4 Discuss how Landsat TM is used to relate unique spectral signatures to specific minerals. What is the difference in ratio between montmorillonite and illite mineral reflectances at 1 and 1.4 microns? What does this show about the choice of wavelength for discrimination? Why are some minerals more valuable than others?

7.5 With respect to figure 7.1, suggest how Landsat satellite imagery may help to discriminate between mineral group 1 with high redness and mineral group 3?

7.6 Three deposits containing Iron (Fe) and Manganese (Mn) are given: a) 40% Fe 0.1% Mn, b) 1% Fe 40% Mn and c) 5% Fe 0.01% Mn. What are the respective Fe/Mn ratios for each? If the Earth's crust has an Fe/Mn ratio = 0.53, which deposit comes from the Earth's crust?

7.7 Explain the geological oil exploration remote sensing process.

7.8 Consider figure 7.2 and indicate where on the soil triangle a 20/40/40 mixture of sand/silt/clay occurs. What sandy loam values are typical, as distinct from silty clay?

7.9 Explain the value of SMOS and how satellite signals depend upon moisture and salinity.

7.10 Explain how SAR radar systems achieve improved subsurface penetration. What applications can space-based SAR systems be used for?

References

[7.1] M Hansen (2016), www.usgs.gov/center-news/landsat-eyes-help-guard-world-s-forests

[7.2] 'Application of high-resolution satellite imagery to environmental assessment of mining impact in West Papua', CR Lavers, J Mazower and S Grig, Proceedings of RSPSoc Annual Conference 2010.

[7.3] 'Use of high resolution NDVI and Temporal satellite Imagery in Change Detection for Assessment of Mining, Construction, Environmental and Ethnic Impact', CR Lavers and J Mazower, Proceedings of RSPSoc Annual Conference 2013.

[7.4] 'Some Urban Measurements from Landsat Data', Bruce Forster, *Photogrammetric Engineering and Remote Sensing*, Vol. 49, No. 12 (December 1983), pp.1693–1707.

[7.5] 'Automated Mapping of Mineral Groups and Green Vegetation from Landsat thematic Mapper Imagery with an Example from the San Juan Mountains, Colorado', BW Rockwell, US Geological Survey Scientific Investigations Map 3252, 25-p. pamphlet, 1 map sheet, scale 1:325,000 (2013) pubs.usgs.gov/sim/3252/

[7.6] 'Observer Based Comparative analysis between thermal, visible and Near Infrared images recorded at Dartmouth Castle, England', CR Lavers et al., Proceedings of RSPS04.

[7.7] 'Saving Africa's Soils: Science and Technology for Improved Soil Management in Africa', MJ Swift, KD Shepherd (Eds), 2007. World Agroforestry Centre, Nairobi.

[7.8] www.africasoils.net

[7.9] en.wikipedia.org/wiki/Soil_texture

[7.10] speclab.cr.usgs.gov/PAPERS.refl-mrs/refl4.html

[7.11] glcf.umd.edu/data/modis/index.shtml

[7.12] glcf.umd.edu/data/lc/

[7.13] 'Fine spatial resolution satellite observations for monitoring human rights and environmental protection issues in Zimbabwe', CR Lavers, SENSED (RSPSoc Newsletter), No. 37, July 2010, pp.6–7.

[7.14] 'Monitoring phenological cycles of desert ecosystems using NDVI and LST data derived from NOAA-AVHRR imagery', G Dall'Olmo and A Karnieli, *International Journal of Remote Sensing*, Vol. 23, No. 19 (2002), pp.4055–4071.

[7.15] 'Penetration of 0.1 GHz to 1.5 GHz electromagnetic waves into the earth surface for remote sensing applications', PK Kadaba, Proceedings of the Southeast Region 3 Conference, Clemson, S.C., April 5–7 1976. (A76-47201 24-99) New York, Institute of Electrical and Electronics Engineers, Inc., pp.48–50.

[7.16] 'Vegetation and Population Density in Urban and Suburban Areas in the U.S.A.', F Pozzi and C Small, Proceedings of the Third International Symposium of Remote Sensing of Urban Areas, 11–13 June 2002, Istanbul, Turkey, pp.489–496.

[7.17] 'Epidemiological and Viral Genomic Sequence Analysis of the 2014 Ebola Outbreak Reveals Clustered Transmission', JP Townsend et al., *Clinical Infectious Diseases*, 60(7) 1 April 2015, pp.1079–1082.

[7.18] rapidfire.sc.i.gsfc.nasa.gov

[7.19] unitar.org/unosat/

[7.20] www.aaas.org/page/high-resolution-satellite-imagery-ordering-and-analysis-handbook

8
Atmospheric Applications

'I was hit with the realisation that this delicate layer of atmosphere is all that protects every living thing on Earth from perishing in the harshness of space' Ron Garan, US Astronaut

8.1 Atmospheric remote sensing applications

The atmosphere is perhaps the hardest region to measure and consider conceptually, as most of the atmosphere's properties are invisible to the eye even in the visible, unlike imagery of Earth, sea or cryosphere. There is no continuous 'solid' surface to provide clear reflections. Gaseous content is important to characterise vital 'health' information on the state of the planet, monitoring levels of gaseous species and particulates. Additional difficulties arise from the fact that the atmosphere changes continuously and we are measuring huge sample volumes. But why should we study these atmospheric gases? There are several good reasons to start with: global warming due to increased abundance of greenhouse gases, destruction of stratospheric ozone due to increased N_2O, increased tropospheric ozone due to vehicle and anthropogenic contributions, changes in the density of tropospheric clouds, aerosols in the stratosphere such as acid rain, natural wildfires and man-made burning.

We are interested in monitoring O_3, NO_2, BrO, CH_4, H_2O vapour, HCHO, SO_2 and other gases besides. CH_4 (methane) is an effective greenhouse gas and its concentration in the atmosphere has increased over recent decades. Deterioration in air quality happens when high levels of some aerosols enter the troposphere. Global emissions of human activities were estimated at 41 billion tonnes in 2017. The global cost of poor air quality is estimated to be in excess of US$3 trillion per annum (OECD, 2016). People are responsible for 67 per cent of total methane emission, running at about 600Tg yr^{-1}, the main sources being dairy farming, rice fields, waste management and natural gas. It is important to target the most highly polluted areas with high population densities. Direct access to Copernicus air quality data provides 23km pixel resolution, not adequate for useful city-wide analysis, but it is hoped that satellite-based solutions will provide 7km pixel resolution over UK cities on an hourly basis by 2020. Satellite-derived particulate matter (PM2.5) measurements derived from natural sources and traffic (particularly diesel cars) are becoming very important. In 2015 the Global Burden of Disease Study concluded that PM2.5 (particulates much less than 2.5 microns) was the fifth leading cause of deaths, accounting for 4.2 million deaths per annum.

Satellites play a significant role in providing information on topics such as atmospheric chemistry (Aira, ACE-Odin, MIPAS-Envisat), meteorology (Nimbus-D, CALIPSO, CloudSat, and AIRS-Aqua), atmospheric ozone and pollution in the troposphere using MOPITT-Terra, and ozone in the stratosphere, thermosphere and mesosphere with UARS (the Upper Atmosphere Research Satellite), TOMS-Nimbus, GOME-ERS-2, OMI-Aura and GOMOS-Envisat.

Satellite and terrestrial sensors have different spectral and spatial resolutions. Table 8.1 shows typical satellite-based atmospheric systems with spectral and spatial resolutions.

Instrument	Spectral range microns	Spectral resolution (nm)	Horizontal resolution (km)	Vertical resolution (km)
SCIAMACHY	0.25–2.0	0.25–0.4	30 × 60	–
GOME-2	0.24–0.79	0.24–0.53	40 × 40	–
OMI	0.27–0.5	0.45–1.0	13 × 24	–
TES	3.2–15.4	0.06 cm^{-1}	5.3 × 8.5	3–5
MOPITT	2.3–4.67	(0.1 cm^{-1})	22 × 22	3–5
PanFTS	0.27–14.0	(0.05 cm^{-1})	7 × 7	2–5

Table 8.1: *Spectral and spatial resolutions for different sensors.*

Four NASA satellites – Aqua, Aura, CloudSat and CALIPSO – and the French Space Agency's PARASOL make up the so-called satellite 'A-Train'. The limb sounder on the A-Train's Aura monitors cloud CO, which usually comes from power plants or agricultural fires. CO is stable up to two months in the atmosphere, while methane is stable up to five years and CO_2 for 15 years.

Atmospheric greenhouse gases concentration can be retrieved from spectra obtained with ADEOS (via interferometric monitoring of greenhouse gases) and MOPITT. Aura was designed to answer questions about life-sustaining changes in the atmosphere. In addition, Envisat carried Global Ozone Monitoring by

Occultation of Stars (GOMOS), Michelson Interferometric Passive Atmospheric Sounder (MIPAS) and Scanning Imaging Absorption Spectrometer for Atmospheric Cartography (SCIAMACHY). The Global Ozone Monitoring Experiment (GOME) on board ERS-2 provided complementary information about atmospheric trace gases.

We introduced the atmosphere in Chapter 5, and will look at sensing as a function of height, broken into the divisions of **troposphere:** 0–15km, **stratosphere:** 15–50km, **mesosphere:** 50–80km, **thermosphere**, usually above 80km altitude, and **exosphere**, 500km and above, ranging from terrestrial to satellite-based sensors. The history of atmospheric remote measurements started with observation missions focused on Earth's radiation budget because the radiative exchange between Earth and space helps evaluate the driving mechanisms of terrestrial weather and climate (figure 8.1).

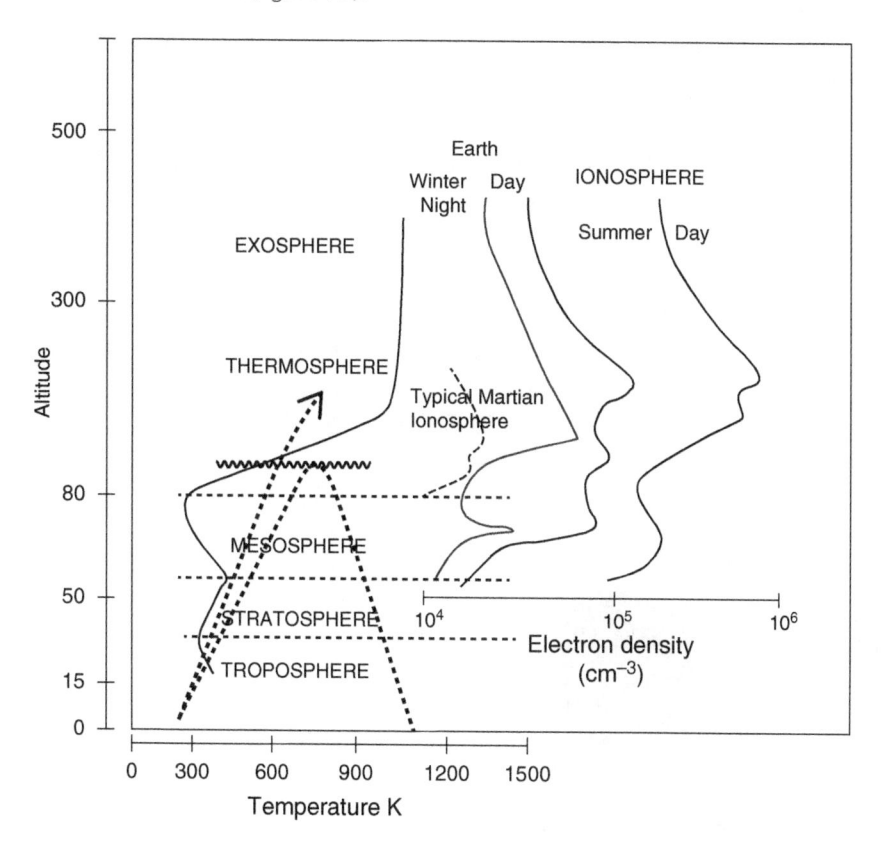

Figure 8.1: *Typical ionospheric structure, Earth winter night and day, and summer electron densities. Typical Martian electron densities, transmitted and totally internally reflected wave from ionospheric layer.*

8.2 Measurement geometries

There are three geometries commonly used for measuring aerosols in the upper atmosphere and elsewhere (figure 8.2):

(1) *Nadir* Downward-looking

(2) *Limb* Looking through the edge of the atmosphere with cold space as a background over a range of elevations above Earth's surface with solar illumination behind.

(3) *Occultation* With this technique, a satellite looks back towards the sun through an extended atmospheric path, providing a much longer path through it than the limb method. It is easier to determine the altitudes of tropospheric observed aerosols. The simplest way to view occultation is to think of a space probe on the opposite side of a distant planet that needs to send a known radio frequency signal back to Earth. As the signal grazes the planet's limb, its radio signal is occulted – that is, it is refracted or bent. If, on the other hand, the planet has no atmosphere, the radio signal received on Earth will have travelled in a straight line only. As the radio signal passes through the atmosphere, the Doppler shift received from a direct path and a refracted path are different. Planetary radio occultation was first observed using Mariner IV while in orbit around Mars in July 1965. Radio occultation from Earth's outer atmosphere can be conducted using dedicated research satellites, but occultation also affects the performance of many other types of satellites including modern GPS systems such as Navstar GPS, GLONASS and Galileo, providing some 80 plus satellites to compare signals.

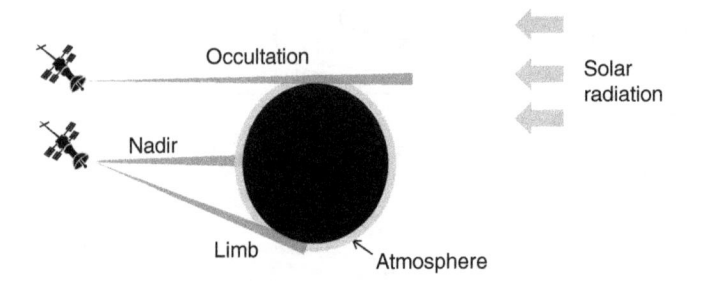

Figure 8.2: *Three common satellite atmospheric constituent geometries: nadir, or downward-looking, limb and occultation.*

Many measurements are recorded from terrestrial, atmospheric and space-based sensors to monitor gases and to model the processes taking place. To these ends, the US Air Force Research Laboratory and Spectral Sciences Inc. developed and maintain a computer code called MODTRAN® (MODerate resolution atmospheric TRANsmission). MODTRAN is used worldwide by research scientists in government agencies, commercial organisations and educational institutions to predict and analyse optical measurements through the atmosphere. The code is embedded in operational and research sensor and data processing systems, particularly those involving removal of atmospheric effects, commonly referred to as atmospheric correction, in remotely sensed multi- and hyperspectral imaging (MSI and HSI). The current version is MODTRAN6, which computes LOS radiances by integrating through a stratified spherical atmosphere, including the effects of light refraction. Further details are found elsewhere [8.1].

8.3 Atmospheric layer sensing

8.3.1 Exosphere

This region begins 500 to 1,000km above Earth's surface and concerns space weather effects largely unprotected from Earth's magnetosphere and denser atmosphere. Satellites such as ACE, SOHO and Cluster have been used to monitor the magnetosphere, besides others like IBEX, Ionozond and THEMIS. Such satellites can monitor total solar irradiance, galactic cosmic ray, neutron count, geomagnetic aa-index, auroral incidence, sunspot number and cosmogenic ^{10}B concentration; ^{10}B has the longest sunspot data set (c.1600) including the Maunder minimum (1645–1715), aurorae (1785), aa-index since 1850, neutrons c.1850 and irradiance c.1990. Solar storms are hazardous to modern monitoring satellites, like a wind anemometer is vulnerable within a hurricane!

ACE is the Advanced Composition Explorer, a NASA Solar mission to study matter comprising energetic particles from the solar wind, the interplanetary medium and other sources. Real-time data from ACE is used by the NOAA Space Weather Prediction Center to improve solar storm forecasts and warnings. ACE was launched in 1997, close to the L1 Lagrangian point, 1.5 million kilometres from Earth. As of 2018, the spacecraft is still operating well, and projected to have enough propellant to maintain orbit until 2024! The satellite has examined blast waves across the sun's atmosphere, the Earth's magnetosphere and geomagnetic storms [8.2].

SOHO, the Solar and Heliospheric Observatory project, is a co-operative effort between ESA and NASA. SOHO was designed to study the sun's internal structure, its extensive outer atmosphere and the solar wind – the stream of highly ionised gas that blows continuously outward through the solar system. SOHO was launched in 1995 and meant to operate until 1998. It was so successful ESA and NASA decided to prolong its life several times [8.3].

Cluster II is an ESA space mission, with NASA participation, to study Earth's magnetosphere over nearly two solar cycles. The mission is composed of four identical spacecraft flying in a tetrahedral formation. The four Cluster II spacecraft were successfully launched in pairs in 2000 from Baikonur, Kazakhstan. In February 2011, Cluster II celebrated ten years of successful scientific operations in space. The mission was extended until December 2018. China National Space Administration/ESA Double Star mission operated alongside Cluster II from 2004 to 2007. These satellites monitor magnetic field and Coronal Mass Ejections (CMEs) from the solar surface. Images of the Earth-facing side of the sun on the Helioseismic and Magnetic Imager (HMI) on NASA's Solar Dynamics Observatory monitor strong solar flares and CMEs [8.4].

The exosphere contains the Van Allen belts, an inner belt of peak ionisation 3,000km above Earth and an outer belt 12,000–22,000km above Earth. However, in 2013 the Van Allen probes showed there is evidence of another, unknown radiation belt. Ionisation creates aurorae over the polar environments and is responsible for the South Atlantic Anomaly (SSA), which affects satellite performance. Space weather provides additional forces on satellites, meaning they deviate from expected positions. GPS satellite constellations measured the position of another satellite subject to such forces, Jason-1, which itself measured distance down to the sea surface. Space weather should concern us all today, given the many space-based GPS geo-positional applications that we rely on. NASA's Van Allen Probes are two spacecraft carrying identical instruments. Both operate together in a highly elliptical orbit, as close as 375 miles, and out to 20,000 miles above Earth, travelling through the belts. A Radiation Belt Storm Probe Ion Composition Experiment (RBSPICE) measures its composition – protons, helium and oxygen – and ring current pressure. These factors impact the shape of the magnetosphere's magnetic fields. Scientists use this information to understand how and why ring current changes during geomagnetic storms and how this, in turn, affects the particles in the belts [8.5].

The probe also carried a Relativistic Proton Spectrometer (RPS) to measure the protons' energy and how their intensity varies over time, to understand how the inner belt changes in response to solar activity. The Electric Field and Waves Suite (EFW) measures electric fields to understand the processes providing energy to the particles; some fields last milliseconds and others hours. The Electric and Magnetic Field Instrument Suite and Integrated Science (EMFISIS) payload measures electric and magnetic field components in three directions, gathering electric field information as well as measurement of various particles (ECT) alongside the monitoring of Helium Oxygen Proton Electron levels (HOPE), the Magnetic Electron Ion Spectrometer (MagEIS) and the Relativistic Electron Proton Telescope (REPT).

Swarm satellites Swarm A–C cross the edge of the exosphere and ionosphere. High-precision and high-resolution measurements of the strength, direction and variations of Earth's magnetic field and electric field provide data for modelling the geomagnetic field and its interactions, building on the earlier Ørsted and CHAllenging Mini-satellite Payload (CHAMP) missions. This three-satellite constellation has shown there is a direct link between GPS blackouts of LEO satellites and ionospheric thunderstorms. During the first two years of Swarm's operation, their GPS connection failed 166 times. High-resolution observations from the satellites link these outages to ionospheric thunderstorms 300–600km up in Earth's atmosphere [8.6]. Space weather and environment have a large impact on the orbits, clocks and attitudes of GPS and GNSS satellites due to highly energetic particles and high frequency electromagnetic energy.

8.3.2 Ionospheric measurements

This region covers both the mesosphere and thermosphere, which we will look at here from both a terrestrial and non-terrestrial measurement viewpoint.

8.3.2.1 Terrestrial measurements of the ionosphere

This is the region of the atmosphere where ion and electron density reaches values that influence refraction of radio waves between 3kHz and 30MHz. Ionospheric plasma is mainly due to photo-ionisation of atmospheric gases caused by UV and X-rays from the sun, and is a very low-density plasma immersed in the geomagnetic field. Guglielmo Marconi was the first to provide experimental proof of the existence of the ionosphere, proposed during the 19th century by Balfour Stewart and Arthur Schuster.

The ionosphere's vertical structure was described during the 1920s with systematic experiments and theoretical studies by Edward Appleton, aided by the technological developments of Americans Breit and Tuve. In the same period, the name 'ionosphere' was coined after a discussion between Appleton and Robert

Watson-Watt, the inventor of radar. The principal method used to investigate the terrestrial ionosphere was vertical sounding, using basic radar techniques, to detect electron density in the plasma as a function of height, scanning transmitting frequency from 1–20MHz and measuring any echo time delay.

Starting from 1930, a network of ionospheric vertical stations was expanded, contributing to a better knowledge of the ionosphere. When an electromagnetic wave penetrates vertically in the ionospheric plasma, total internal reflection occurs at a level where the refractive index becomes zero for the incident frequency wave.

According to the ionospheric theory, the real part of refractive index:

$$n_p = \sqrt{1 - \left[\frac{f_p}{f}\right]^2} \qquad \text{(eq 8.1)}$$

dependent on the plasma frequency f_p given by:

$$f_p = \sqrt{\frac{Nq^2}{4\pi^2 \varepsilon_0 m}} \qquad \text{(eq 8.2),}$$

where N is electron density, q and m are the charge and mass of the electron respectively and f the incident frequency. Reflection in the ionosphere occurs when the incident frequency $f = f_p$.

Simplistically, frequencies below the minimum required for internal reflection are heavily absorbed, while frequencies above the minimum up to the critical frequency are internally reflected by the weakly ionised plasma of gases, returning with increasing delay times until the critical frequency is reached. At this point ionisation cannot refract waves sufficiently to turn them back to Earth, and waves pass into space. The maximum electron density N_m corresponds to the maximum vertically reflected incident frequency, and occurs at the critical frequency f_c.

$$Nmax = 1.24 \times 10^{10} f_c^2 \qquad \text{(eq 8.3)}$$

where Nmax and f_c are expressed in electrons m^{-3} and MHz respectively.

An ionogram (figure 8.3) is produced by an ionosonde, which shows the time between transmission and the received echo from the ionospheric layer. Delay is proportional to altitude as a function of frequency. Since signals travel more slowly in the ionospheric plasma than in free space, above the minimum frequency for layer TIR, the height observed exceeds the real height reflection and is called a virtual height. If the frequency increases, virtual height also increases, but more rapidly than the real height. When the level of maximum electron density in the layer is reached, the delay time becomes infinite, and waves are not returned any longer and virtual height becomes infinite. This frequency is the layer's critical frequency.

Consider radio waves of variable frequency transmitted towards the atmosphere. The ordinary mode (a 'cold' plasma with no magnetic field) of the transmitted wave reflects from the ionosphere when the transmitted frequency matches the electron plasma frequency – that is,

$$f_p = \sqrt{81 N_{max}} \quad \textbf{(eq 8.4)},$$

where f_p depends on the reflection height, with a time delay according to the relation:

$$\Delta t = \frac{2h}{c} \quad \textbf{(eq 8.5)}$$

The experimental data plot virtual reflection height versus transmitted frequency on the ionogram.

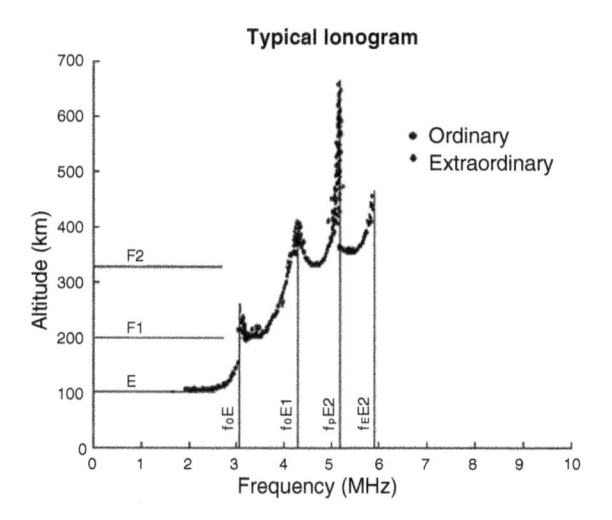

Figure 8.3: *Typical ionogram.*

The task of ionospheric stations is to monitor the ionosphere above each station, obtain data to evaluate long-term changes, study ionospheric phenomena and determine the ionosphere's global morphology. The key layers are the E, F1 and F2 layers. Further refinements include development of oblique ionospheric sounding. Detail about the E layer (ionised molecular oxygen) and F1 and F2 layers (isotopes of nitrogen) are discussed elsewhere [8.7]. Ionospheric products are available online [8.8]. A global ionospheric network exists, providing accurate specifications of electron density in the Earth's ionosphere at over 60 locations worldwide.

8.3.2.2 Thermosphere: space-based methods

Temperatures here depend on solar activity and can rise to 2000°C. Radiation causes atmospheric particles to become electrically charged, enabling radio waves to refract considerable distances beyond the visible horizon. The dynamics of the thermosphere are dominated by atmospheric tides, driven by diurnal heating. Atmospheric waves dissipate above this level because of collisions between neutral gas and the ionospheric plasma. The thermosphere is still monitored using routine ionospheric observations as well as satellites such as TIMED (Thermosphere, Ionosphere, Mesosphere Energetics and Dynamics). The International Space Station also orbits within the thermosphere, typically 330–435km, providing data on its surroundings.

8.3.2.3 Mesosphere space-based methods

This is the layer of Earth's atmosphere directly above the stratosphere and below the mesopause. The mesosphere's upper boundary can be the coldest naturally occurring place on Earth with temperatures below 130K recorded between 50 and 100km. Noctilucent clouds are located in the mesosphere. The upper mesosphere is also the region known as the D-layer. The mesosphere is a difficult region to sense as it lies *above* altitude records for aircraft (about 15km), with only the lowest few kilometres accessible to balloons. Meanwhile, the mesosphere is *below* the minimum altitude for orbital spacecraft due to high atmospheric drag (see Chapter 9). It is only through use of sounding rockets that data has been obtained in this region apart from ionospheric RF measurements. Some large balloons for instruments for submillimetre astronomy have been achieved up to 30km. Lidar is an obvious choice method for any platform, terrestrial or space-based, for sensing this region as laser wavelengths can be chosen to achieve the required operational range.

Different sensors can bring complementary information. However, rockets provide less data than lidar, at higher cost. Radar and satellite systems generally provide low resolution, deteriorating with distance from Earth's surface and atmosphere, while radiosondes are limited in their data provision and subject to wind direction! Lidar can provide high spatial and temporal resolutions and cover heights above 100km, providing temperature, pressure humidity, wind, trace gases detection, aerosols and clouds. Ground-based lidar and airborne lidar applications exist, including remotely operated UAVs using DIfferential Absorption Lidar (DIAL) systems. Space-based applications were available from Aerosol Cloud Lidar and ADM-Aeolus (2018). Raman lidar was used to measure water vapour density at night over Xi'an, China. Non-laser wind profiling systems operate at 300–500GHz, typically with the NPN Network [8.9]. Terrestrial UK air quality mapping is undertaken by the UK government and various universities in collaboration with the UK Space Agency.

8.3.3 Stratosphere satellite-based sensing

Some of the first research satellite work involved sounding the upper troposphere and lower stratosphere in two key areas, starting with use of *nadir-sounding* stratospheric ozone profiling: NASA BUV (1970), SBUV (1978), SBUV/2 (1985–present) and SSBUV (1989–96). This was extended into the troposphere by ESA GOME-1 (1995) and SCIA (2002). GOME helped map ozone holes in the Antarctic, which develop in the southern hemisphere spring, and the Arctic. Long-term ozone monitoring is vital to provide information to understand the processes involved and identify long-term trends.

The second approach is *limb sounding*, which began with SAGE-I–III (1979, 1984, 2001) providing stratospheric ozone profile trends, H_2O and O_3 profiling, pioneered by Nimbus 6 LRIR (1975), and trace gases by Nimbus 7 LIMS and SAMS (1978), others including UARS and MLS (1991), and below the tropopause by Envisat MIPAS (2002) as well as Aura MLS. HIRDLS was the first to observe fine vertical structure with 1km vertical resolution.

Emission limb sounding has some advantages, as none of the emitted radiation seen by the satellite originates below the tangent height, due to the geometry. More emitted gas is encountered along grazing paths than on straight ones, increasing sensitivity to low gas concentrations. However, it cannot be used in the troposphere because the atmosphere is too optically thick.

MIPAS was used to monitor volatile organic compounds (VOCs) such as C_2H_2, C_2H_6 and HCOOH, providing global distributions of acetone in the upper troposphere [8.10]. The space shuttle also played a role with ATMOS (1985, 1992–94) and MAS (1992–94), as has the ISS with SMILES (2009).

Aura Microwave Limb Sounder (MLS) detected unprecedented Arctic ozone loss in 2011. Unusually prolonged cold conditions in the spring 2011 Arctic stratosphere promoted levels of chlorine activation and chemical ozone loss never before observed in the Arctic, comparable to those in Antarctic winters.

Aura's observation of Arctic vortex chlorine monoxide ClO and O_3 at around 18km altitude was useful for convective water vapour measurements and for monitoring North American and Asian monsoon events. Carbon monoxide is a long-lived trace gas that is a primary component of biomass burning and emitted by other anthropogenic processes. OSIRIS provides volcanic background aerosol data in the lower stratosphere.

Nadir-sounding operational polar orbiting systems including MetOp-A–C, IASI and GOME-2 (2006–20) provided statistical agreement with ozone-sondes sampled worldwide, as well as pressure data up to 20km altitude at sub km resolution. Other nadir systems include Suomi NPP, OMPS (2011, 2017), Sentinel-5P (2016–21) and MetOp-SG IASI-NG. Altitude mapping of CH_4 ppm from Sentinel-5 (2020–30) will provide co-located IR and SWIR observations.

SCISAT ACE (2003–present), Odin-SMR and Osiris (2001–present) provide 3D sounding, less than 1km vertical CO and volcanic aerosol sampling in the lower stratosphere. IR and mm-wave limb sounders provide complementary information – that is, target trace gases are provided by IR wave CH_4, organic compounds and nitrogen oxides, while mm-wave provides CO, biomass burning indicators and halogens. IR is controlled by cloud while mm systems are influenced largely by water vapour. Sensors are sensitive to cirrus particle size, IR to less than 100 microns while mm-wave systems provide sensitivity above 100 microns.

8.3.4 Troposphere

8.3.4.1 Terrestrial measurements of the atmosphere

Ground-based networks are used extensively for the atmosphere, including the Network for the Detection of Atmospheric Composition change (NDACC) and the Global Atmosphere Watch (GAW). NDAA regularly measures O_3 total column with

UV/visible spectrometers and DIAL profiling (0–60km), ozone-sondes (0–35km), ClO with microwave radiometers (12–65km), H_2O profile Raman lidars (15–35km), microwave radiometers (40–80km) and aerosol distributions with backscatter lidars and sondes (0–30km). The Total Carbon Column Observing Network (TCCON) is a global network of ground-based instruments designed to measure CO_2, CO, CH_4, N_2O and other gases. TCCON instrumentation includes Fourier transform spectrometer covering the range 3,900–15,500cm^{-1} simultaneously. Spectral fitting and least squares data processing and profile scaling algorithms can retrieve specific gases such as those above. GAW provides a systematic global monitoring of the atmosphere's chemical composition, stratospheric and tropospheric ozone, UV radiation, greenhouse gases, synthetic greenhouse gases (CFCs, SF6 etc), aerosols, reactive gases and natural radionuclides.

Lidar is a common way of measuring stratospheric and tropospheric constituents, similar in operation to radar, and can be thought of as laser radar. Lidar transmits at higher optical frequencies than radar in the UV, visible or IR. For some lidar applications, more than one laser is used, such as CW and pulsed lasers. The detection system records scattered light at fixed time intervals.

Range bins $\Delta z = c \Delta t/2$ (**eq 8.6**),

so if a laser points up vertically with each range bin lasting 160ns, the height of each bin is 24m.

So the question is, how can lidar measure atmospheric aerosols?

One common method is DIAL, which stands for DIfferential Absorption Lidar. Absorption of light by ozone in the atmosphere is different at different wavelengths. Ozone absorption is broad, the laser is tuned between spectral regions of high and low ozone absorption. The different absorption of light at multiple wavelengths determines ozone content, such as in the Antarctic ozone hole. The LITE system has been used to detect clouds and aerosols from space. LITE stands for the Lidar In-space Technology Experiment, and flew on shuttle mission STS-64 in 1994.

An advanced wind lidar, Giant Aperture Lidar Experiment (GALE), measures wind, temperature and waves using resonance fluorescence scattering due to sodium present in our atmosphere. With resonance fluorescence scattering, sodium atoms are illuminated at 589nm, which become excited and radiate light. By adjusting the light wavelength, the shift of spectral line from its central wavelength can be measured, such as with the large aperture Starfire Telescope. It is regularly used to monitor wind changes from 40ms^{-1} W to 80ms^{-1} E, around 90–97km altitude. Wind lidar at tropospheric levels has provided data for the 2008 Olympics in China and to aviation for ash detection, as well as air pollution detection for gases such as NOx, SO_2, O_3, CO and CH_4.

Ultraviolet laser sounding of the troposphere and lower stratosphere is also possible. Measurements made with ground-based ultraviolet lasers operate between 297 and 308nm, from ground level to 20km, particularly for contributions of absorption by minor constituents. There is considerable recent concern about the biological effects of increased solar UV radiation at ground level associated with reduction in atmospheric ozone concentration. Wavelengths between 300 and 310nm are likely to produce erythema (sunburn) and skin cancer. The effects of aerosols, clouds, non-absorbing haze or fog and other absorbing atmospheric gases need to be considered alongside the effects of ozone and Rayleigh scattering when examining the penetration of these radiations to Earth's surface. Except for the Junge layer of sulphur-rich aerosols (12–22km) and cirrus clouds, scattering and absorption sources are mostly in the lower troposphere, below about 5km [8.11]. It is also possible to detect stratospheric N_2O_5 with IR remote sounding.

Measurements Of Humidity in the Atmosphere and Validation Experiments (MOHAVE I, II and 2009), operated by JPL 2006–09, evaluated the capability of three Raman lidars dedicated to water vapour measurements in the upper troposphere and lower stratosphere, besides tropospheric and stratospheric ozone and temperature measurements. Over 200 hours of lidar measurements were compared with balloon measurements from flights and radiosondes [8.12].

Ozone monitoring in the troposphere and lower stratosphere has been performed on an operational basis using a ground-based lidar station set up in 1980 at the Haute-Provence Observatory (France). Results obtained during several experiments (1980–81) showed ozone variability in the troposphere and lower stratosphere, ozone exchange between these two regions, and the evaluated tropospheric ozone budget.

8.3.4.2 Space-based tropospheric measurements

The Along Track Scanning Radiometer ATSR on board ERS-2 had four visible and mid IR channels, centred at 0.55, 0.67, 0.87 and 1.6 microns, and three thermal IR channels, centred at 3.7, 11.0 and 12.0 microns. These produce a double forward view at 55° off-nadir of the same surface at 1km resolution. The satellite cycle was 35 days and data enabled the creation of the world's first fire atlas (1997–98). This data compared biomass burning sources with relative aerosol production. ATSR2 satellite imagery has been used to detect ship tracks due to ships modifying cloud microphysics by aiding cloud condensation nuclei to developing or existing clouds [8.13].

CloudSat was an Earth observation satellite that used radar to measure the altitude and properties of clouds, flying in the A-Train with Aura and CALIPSO, a 94-GHz nadir-looking radar that measures cloud backscattered power as a function of distance from the radar.

The Tropical Rainfall Measuring Mission (TRMM) operated by JAXA (Japan Aerospace Exploration Agency)/NASA in LEO was a joint mission to monitor tropical rainfall. It operated for 17 years and re-entered Earth's atmosphere in June 2015. Tropical precipitation is a difficult parameter to measure, due to the large temporal and spatial variations that occur. However, understanding tropical precipitation is important for weather and climate prediction, as precipitation contains 75 per cent of the energy that drives atmospheric wind circulation and helps predict the onset of El Niño and evaluate diurnal variability of tropical rainfall measurements. The satellite itself sat in a fairly circular orbit of 174–176km for most of its working life. TRMM carried a Microwave Imager (TMI), Visible and InfraRed Scanner (VIRS) and a Clouds and the Earth's Radiant Energy Sensor (CERES).

The latest Sentinel-5 Precursor Tropomi instrument is dedicated to monitoring air pollution. Tropomi stands for TROPOspheric Monitoring Instrument, and is a spectrometer sensing UV, visible, NIR and short wavelength IR to monitor ozone, methane, formaldehyde or HCHO (forest fires and wood processing), aerosols, CO from volcanic activity, NO_2 (roads and traffic) and SO_2. It extends the capabilities of Aura's OMI sensor and Envisat's SCIAMACHY instrument. Tropomi takes measurements every second over an area 260km wide by 7km long at 7 × 7km resolution. Light is separated into different wavelengths using grating spectrometers, measured with four different detectors. The UV spectrometer has a range of 270–320nm, the visible light spectrometer covers 310–500, NIR 675–775 and SWIR 2305–2385nm. SCIAMACHY (SCanning Imaging Absorption spectroMeter for Atmospheric CHartographY) was one of ten instruments aboard Envisat, designed to measure sunlight transmitted and reflected and scattered wavelength dependent radiation from Earth's atmosphere or surface in the UV to NIR bands at moderate spectral resolution. The mission ended in May 2012. SCIAMACHY was important in exploring the chemistry and physics of

Earth's atmosphere (troposphere, stratosphere and mesosphere) and changes resulting from anthropogenic or natural phenomena. The Total Ozone Mapping Spectrometer (TOMS) is a NASA satellite instrument for mapping ozone. Four TOMs imaged a range of important atmospheric aspects – ash, SO_2 and other aerosols – from Mount Pinatubo in 1991, and the largest Antarctic ozone hole recorded (September 2006).

Another tropospheric monitoring instrument, MOPITT (Measurements Of Pollution In The Troposphere) is a payload scientific instrument launched into Earth orbit by NASA on board the Terra satellite in 1999 (figure 8.4, see plate section). It monitored changes in pollution patterns and its effect in the lower Earth atmosphere, mapping global CO. It is a nadir-sounding instrument measuring upwelled IR radiation at 4.7 microns and 2.2–2.4 microns. It uses correlation spectroscopy to calculate total column observation and CO profiles in the lower atmosphere.

8.4 Satellite validation principles

Remote sensing data is affected by things like instrument degradation, which affects retrieval of the correct data parameters, as well as calibration uncertainties and retrieval model errors. Validation is necessary to confirm results derived from theoretical and laboratory-based studies. Quantitative validation takes into account satellite and other related data sources. Additionally, Earth observation doesn't provide a direct measure of the atmospheric parameters, being usually an algorithm-related correlation, and is highly dependent on Earth radiance and solar irradiance measurements and on horizontal resolution: latitude and time. The measurement and retrieval algorithm approach is characterised by 'weighting functions' determining altitude range, sensitivity, and the vertical resolution of retrieved information.

Validation characterises the real measurement information content. Satellite and ground-based data give a smoothed/sampled/truncated perception of real ozone levels. Atmospheric ozone exhibits spatial structures and temporal variability. Validation must combine both a qualitative and quantitative approach.

8.5 Available products

NASA provides an extensive atmospheric data resource called Giovanni, where maps of trace gases can be freely obtained with a resolution of around 100km [8.15]. Sensitivity and resolution depends on the specific chemical's nature and the specific instrument (see table 8.1). A table of typical trace gases, with past and present sensors, is given in table 8.2.

Atmospheric trace gas	Typical past and present sensors
BrO	OMI, MLS, GOME, SCIAMACHY
ClO	MLS
O_3	TOMS, OMI, MLS, SBUV, GOME, SCIAMACHY, GOMOS
Water vapour	SCIAMACHY, GOME, AIRS
NO_2	OMI, GOME,
N_2O_5	MIPAS
CO	MOPITT, SCIAMACHY
CO_2	MOPITT, SCIAMACHY
HCHO	SCIAMACHY, GOME
CH_4	MOPITT, SCIAMACHY, MIPAS

Table 8.2: *Some typical trace gases, past and present sensors.*

Questions

8.1 Use equation (8.1) to find the real part of refractive index if the plasma frequency is 10MHz and the critical frequency transmitted vertically from the transmitter is 15MHz.

8.2 Using equation (8.5), find the time delay for a virtual height of 85km. Explain what is meant by virtual height and why virtual heights from a layer change.

8.3 With equation (8.2), find the plasma frequency if the critical frequency is 12MHz, N the electron density = 10^{11} m^{-3}(MHz)^2 and q and m are the charge and mass of the electron respectively.

8.4 If the critical frequency = 12MHz, what is the maximum electron density?

8.5 Using equations (8.1) and (8.4) for a maximum D region electron density of 3.24×10^4 and frequency = 1.8kHz, find the refractive index of a wave passing through the region.

8.6 Find the critical frequency if reflection takes place at 380km height and maximum ionospheric density n = 0.95 at 15MHz. Derive an expression for the MUF in terms of skip distance D and virtual height h_v.

8.7 For oblique refraction MUF = $\frac{f_c}{\sin E}$ where E is the incident angle, find the MUF possible for a critical frequency = 6MHz and incident angle = 30°.

8.8 An aircraft has a choice to cruise at either 15km or 22km. Which is the better choice and why?

8.9 On Mars, the atmosphere is mainly CO_2, the temperature is 220K and g = 3.7ms^{-2}. What is the Martian atmosphere pressure at 100m height, given the surface pressure is 6mbar?

References

[8.1] modtran.spectral.com/

[8.2] www.srl.caltech.edu/ACE/

[8.3] www.nasa.gov/mission_pages/soho/overview/index.html

[8.4] sci.esa.int/cluster/

[8.5] rbspice.ftecs.com/

[8.6] www.esa.int/Our_Activities/Observing_the_Earth/Swarm/Swarm_reveals_why_satellites_lose_track

[8.7] *Essential Sensing and Telecommunications for Marine Engineering Applications*, C Lavers (Bloomsbury Publishing, London, 2017, ISBN 1472922182).

[8.8] ionos.ingv.it/spaceweather/start.htm

[8.9] climateviewer.org/pollution-and-privacy/atmospheric-sensors-and-emf-sites/maps/noaa-profiler-network-npn-wind-radar/

[8.10] 'Global distributions of acetone in the upper troposphere from MIPAS spectra', DP Moore, JJ Remedios and AM Waterfall, *Atmospheric Chemistry and Physics*, 12 (2012), pp.757–768, https://doi.org/10.5194/acp-12-757-2012

[8.11] 'Ultraviolet laser sounding of the troposphere and lower stratosphere', AJ Gibson and L Thomas, *Nature* Vol. 256 (1975), pp.561–563.

[8.12] tmf.jpl.nasa.gov/tmf-lidar/results/water_vapor.htm

[8.13] 'Detection of ship tracks in ATSR2 satellite imagery', E Campmany et al., *Atmospheric Chemistry and Physics*, 8(4), August 2008, pp.14819–14839.

[8.14] www.asc-csa.gc.ca/eng/satellites/mopitt.asp

[8.15] giovanni.gsfc.nasa.gov

9

Satellite Platforms for Remote Sensing

'The world itself looks cleaner and so much more beautiful. Maybe we can make it that way – the way God intended it to be – by giving everyone, eventually, that new perspective from out in space.' Roger Chaffee, astronaut, who died alongside Gus Grissom and Ed White in the Apollo 1 launch pad fire, 27 January 1967.

9.1 An early history of non-terrestrial platforms

Balloons and kites For decades, scientists have used many remote-sensing instruments on different platforms, with ground-based equipment and those on balloons, aircraft and satellites collecting ozone data, unsurprisingly since British Antarctic Survey scientists first discovered ozone layer changes over the Antarctic with atmospheric sounding equipment carried by balloons. Findings were confirmed by NASA's Total Ozone Mapping Spectrometer (TOMS) on board polar orbiting weather satellites, such as Nimbus. Ozone loss occurs every Antarctic spring, and was first recorded from 1979. Work with balloon sounding equipment enabled scientists to plot a total ozone column over Halley Research Station, near latitude 75° south. Four-metre French Met balloons for radar measurements in the 400–460MHz range are still used today. Balloons rise to a pre-planned height and drift, and have been launched in satellite pre-flight calibration activities from Kiruna, Norway for Envisat sensor validation. ESA's PIROG High Altitude Balloon project sent a 500kg gondola up to 40km altitude to investigate the IR interstellar spectrum. The first balloon aerial photo was taken in 1858 by Gaspard-Félix Tournachon, who ascended 80m, taking a photo over Bièvre; however, kites were introduced surprisingly late, with the first image taken in 1882 by the English meteorologist ED Archibald.

Aircraft A limitation of balloons is dependence on wind direction. However, aircraft can target specific areas at little notice independent of wind direction, permitting accurate weather phenomenon measurement, such as tropical storms. Aircraft reconnaissance remains an irreplaceable source of accurate data on the state and location of hurricanes that threaten the Caribbean and USA. Wind strength and centre location are evaluated on the basis of the geosynchronous satellite images – but are not as precise as penetrating aircraft measurements. A key conclusion of the US Department of Commerce's Natural Disaster Survey Report on Hurricane Hugo (1989) was 'aircraft reconnaissance will remain a necessary tool in forecasting hurricanes until other sensing platforms can provide data fields of equal accuracy', and this is still true today.

9.2 Rocketry

The origin of satellite launchers lies in rocketry. A rocket is a reaction device, working in accordance with Newton's Third Law of Motion: 'For every action there is an equal and opposite reaction.' A rocket engine's reaction discharges gas propellant to makes a rocket fly in the opposite direction. The reaction principle was noted c.360 BC by Aulus Gellius, who designed a steam device; Hero of Alexandria is credited with inventing a more sophisticated version in the first century AD. The first Western claims to rocketry development originate in 1450 with the Italian military engineer Robertus Valturius, who wrote of what may have been rockets dating from AD c.886–911. The first Chinese claims of black powder use – charcoal, sulphur and saltpetre, and firecrackers – date from the Qin (221–207 BC) or Han (206 BC–AD 220) dynasties. North African Arabs developed fire arrows and rockets during the 13th century, described by the Muslim historian Ibn Khaldūn in 1384. Rockets were developed over centuries in various forms, leading to those able to launch satellite vehicles.

As a rule of thumb, the rocket mass delivered into LEO is 10–20 per cent of launch pad mass. Some launchers are shown in table 9.1. Elementary classical mechanics shows a rocket of total initial mass Mi burns a mass M_f of fuel to increase velocity. In the absence of gravitational and frictional forces,

$\Delta V = U \ln(M_i/(M_i - M_f))$ (**eq 9.1**),

where U is the exhaust gas velocity with reference to the rocket. As satellite orbital velocity in LEO is ~7kms^{-1}, and U ~2.4kms^{-1}, an estimate for a rocket reaching LEO must be >95 per cent fuel. If gravity and drag are included, this increases to 97 per cent! Payload represents a small fraction of the remaining 3 per cent and so Single-Stage-To-Orbit (SSTO) rockets place only small masses in orbit. Instead, multi-stage rockets, three or four stages, put payloads of a few tons into LEO and smaller payloads into GEO. The US space shuttle could place ~30 tons into a 400km orbit, and 6 tons into GEO. The Soviet Buran shuttle could put larger payloads into orbit.

Country	Vehicle	Lift capability kg	Launch cost US$
USA Boeing	Space Shuttle	25,000	$550M
French Arianespace	Ariane 5	6,300	Unknown
USA Boeing	Delta 3	3,800	Unknown
China	Long March	1,000–4,500	$20–70M
US (SpaceX)	Falcon 9 Block 5	22,800 to LEO	$49.9–56M
US Lockheed Martin	Titan IV	5,700	$350M
US (SpaceX)	Falcon Heavy	63,800 kg to LEO	Reusable: $90M Expendable: $150M

Table 9.1: *List of some space launching vehicles.*

9.3 What is a satellite?

Satellites are objects in orbit around another body – for example, a natural satellite such as the moon, orbiting Earth. The Soviet Union launched the first artificial satellite, Sputnik, in 1957. Today there are thousands of spacecraft in orbit, serving different purposes: some part of communications networks, like Iridium, or carrying instruments to investigate Earth's ocean surface or atmosphere, like Aqua. Some spacecraft face the sun and monitor it, such as SOHO and ACE, or travel into deep space carrying probes, such as Voyager, or investigate atmospheres and moons of distant planets. There are manned spacecraft such as the International Space Station (ISS), which carry out a range of experiments. Since Sputnik's launch and the dawn of the satellite age, we have dramatically increased knowledge and understanding of the world's weather and climate. The first weather satellite, TIROS 1, was launched in 1960, and cloud pictures first became available with ESS-1 in 1966.

Weather satellites provide a continuous atmospheric picture. A network of just surface and upper air observation stations only detects weather at discrete points, some thousands of miles apart. However, satellite imagery fills these gaps, detecting weather systems otherwise missed. It is critical for maritime users to locate and track tropical storms over the oceans where weather observations are sporadic. Satellites have automatic data collection systems relaying data on environmental factors from weather stations, ocean buoys, ships, aircraft and balloons to ground stations, which, when combined, create a full planetary real-time network of weather conditions.

Most Earth observation satellites are unmanned, carrying instruments but no people. The Space Shuttle was a manned spacecraft, carrying astronauts, scientists and instruments. NASA carried out many space experiments with the shuttle, recording Earth's surface with different sensors. The more important sensors on the shuttle for Earth imaging were a shuttle imaging radar, a metric camera, a large format camera and a Modular Optoelectronic Multispectral Scanner (MOMS). As the Space Shuttle is manned, we will limit discussion of the shuttle and ISS.

9.4 Satellite physics basics

Thanks to the work of space scientists, based on the theoretical understanding established by Sir Isaac Newton in the *Principia* (1687), we know the attractive force F_a on an object mass near Earth is given by:

$$F_a = \frac{\text{object mass } m \times G \times \text{mass of the Earth } M}{\text{distance R to Earth's centre}^2} = \text{object mass } m \times g_0 \quad \textbf{(eq 9.2)}$$

where $g_0 = 9.81 \text{ms}^{-2}$, the acceleration due to gravity at Earth's surface, G is the universal gravitational constant $= 6.67408 \times 10^{-11} \text{m}^3\text{kg}^{-1}\text{s}^{-2}$. As a function of height above Earth, the value of gt is given by:

$$gt = \frac{g_0 R^2}{(R+H)^2} \quad \textbf{(eq 9.3)}$$

where R is Earth's radius (nominally 6,400km) and H is the height above Earth in km. Even at an orbital height of 400km typical of the ISS, acceleration due to gravity is 89 per cent of that at its surface.

An object in orbit around Earth is considered as 'free-falling with style', where the force of gravity equals the centripetal force F_c on the orbiting mass. The value of F_c is given by:

$$F_c = \frac{\text{object mass } m \times \text{orbital velocity } v^2}{\text{distance to the centre of rotation}^2} \quad \textbf{(eq 9.4)}$$

Equating this to the reduced force of gravity, orbital velocity is related to height H, so:

$$\text{orbital velocity} = \sqrt{\left(\frac{g_0 R^2}{R+H}\right)} \quad \text{(eq 9.5)}$$

This environment doesn't provide a region without gravity, but rather a region of reduced gravity or *micro gravity*.

9.4.1 Circular orbit and velocity

Imagine a 100nm high tower, from which we launch a satellite in the horizontal direction. If launch speed is low, the satellite falls under gravity and lands on the Earth. The faster the launch speed, the greater distance it travels from the tower before landing. If a satellite is launched at 8kms^{-1} it doesn't strike the Earth, but stays parallel with it, returning to the top of the tower! This is known as a circular orbit and the velocity as circular velocity.

In circular orbit at Earth's surface, $F = mv^2/R = mg$ **(eq 9.6)**

$v^2/R = g \quad v_{circular} = (gR)^{1/2}$

Now, R~6400km so $v_{circular}$ ~8kms^{-1}

Due to gravity, the circular velocity required to keep spacecraft in orbit decreases as distance from Earth increases.

9.4.2 Elliptical orbits and escape trajectories

If velocity is increased, the circular orbit becomes elliptical and then more eccentric until, at around 11.4kms^{-1}, the satellite breaks from the closed elliptical orbit and becomes a space probe, which is eventually captured by the gravitation pull of another body. The first escape trajectory is parabolic but with increasing speed a series of hyperbolic trajectories follows (figure 9.1).

Since escape velocity reduces with increasing distance from Earth, **what minimum speed must we give it?**

From potential energy considerations, the minimum escape speed at radius R is given by:

$1/2\ mv^2_{escape} = GMm/R$

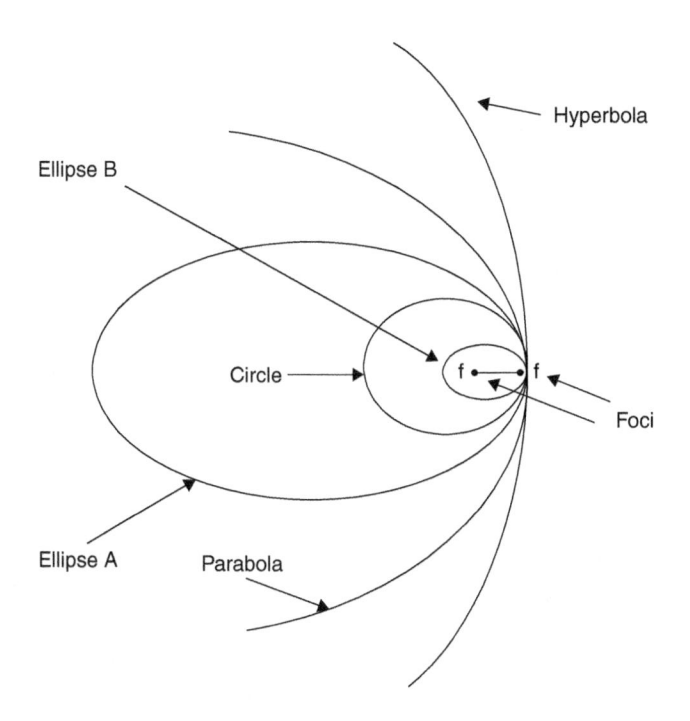

Figure 9.1: *Various orbits.*

$V^2_{escape} = 2GM/R$

Thus: $V_{escape} = (2GM/R)^{1/2}$ **(eq 9.7)**

The gravitational force on a mass m at r = R is equal to mass m times local acceleration due to gravity g, so

$GMm/R^2 = mg$, as $GM/R = g$

Thus: $V_{escape} = (2GR)^{1/2}$

As g » 9.8ms⁻², R » 6400km V_{escape} »11.23kms⁻¹ in **any** direction!

9.4.3 Kepler's laws

In 1609 Kepler introduced three laws of planetary motion, deduced from studying planetary movements around the sun, with data acquired by Tycho Brahe.

First law This states that the plane of an elliptical orbit must pass through the centre of Earth because one of the foci of the ellipse lies at Earth's centre. This law has practical application today in choice of launch sites. Satellites are launched due east to make maximum use of Earth's rotation.

Second law This relates to the satellite speed around an elliptical orbit. It states that the speed of an object around an elliptical orbit changes. At **apogee**, the furthest point from Earth, a satellite has minimum forward speed, while at **perigee**, the nearest point, it has maximum forward speed. The line joining a satellite to Earth will sweep out equal areas in equal times (figure 9.2).

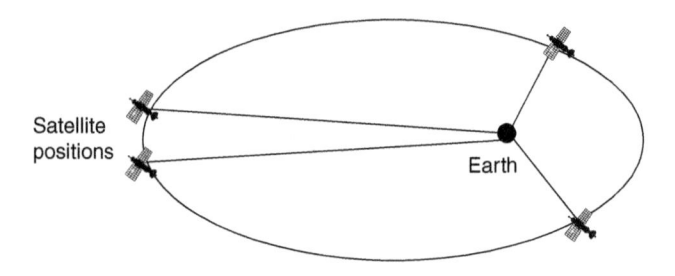

Figure 9.2: *A satellite sweeps out equal areas in equal times.*

Third law This concerns the satellite period, the time taken to complete one Earth orbit. It states that the square of the period is proportional to the cube of the semi-major axis. If two orbits have the small length major axis distance from apogee to perigee, they will have the same period regardless of ellipticity.

Taking a circular orbit, $F = mv^2/R = MmG/R^2$

So $v^2/R = MG/R^2$ but $v = R\omega = R(2\pi f) = 2\pi R/T$

Hence $V^2 = 4\pi^2 R^2/T^2$

And $4\pi^2 R^2/T^2 = GM/R$

So $4\pi^2 R^3/T^2 = GM$ **(eq 9.8)** and $R^3 \propto T^2$, Kepler's third law.

Satellite lifetime is governed by initial apogee and perigee, but perigee is more important. It isn't possible to establish a satellite in circular orbit below 100nm, and below 150nm lifetimes are still very short.

Satellite	Perigee nm	Apogee nm	Orbital lifetime
'Spy' satellite	64	232	15 days
Sputnik 1	122	511	3 months
Transit navigation	590	600	1,000 years
Geosynchronous satellite	19,363	19,363	Indefinite period

Table 9.2: *Apogee and Perigee altitudes for different satellite orbits.*

9.5 Types of satellite

Two types of satellite orbit provide important meteorological and Earth observation data to civilian and military users: polar orbiters and geostationary satellites (figure 9.3, see plate section).

9.5.1 Polar orbiting satellites or Polar Orbiters

Polar Orbiters (PO), such as the Russian Meteor series, are *sun-synchronous*, with orbital paths fixed relative to the sun while Earth appears to rotate below. Typically, PO satellites are 400–900km above Earth's surface. They are relatively near in orbital terms and provide detailed imagery. It takes 90 minutes for a satellite to complete each orbit, passing over the Arctic and Antarctic. Each orbit scans a little further *west*, about 25° on each pass. Thus a different **swath** of Earth's surface is viewed every time. During each pass, satellite instruments scan side to side so each part of Earth's surface is observed at least once every 12 hours by a single satellite, and more frequently at high latitudes where swaths overlap. NOAA PO satellites provide the UK with detailed polar orbiting pictures. Current NOAA 15, 18–19 pass over Britain hours apart, providing daily pictures over the Atlantic and Western Europe, helpful to mariners. The NOAA AMSU microwave imager shows precipitable water, in a long green band sweeping towards the UK, figure 9.4 (see plate section).

9.5.2 Geostationary satellites

Geostationary satellites are Earth synchronous, orbiting at a rate matching Earth's rotation, staying above the same spot on the ground and scanning the same region. Once a satellite dish is installed, there is no need to change its position, as it always points to the satellite. Geostationary orbit was first suggested in 1945 by science fiction writer Arthur C Clarke in the magazine *Wireless World*, from discussion about what to do with old V1 and V2 rockets. From 36,000km above Earth's surface (one-tenth of the way to the moon) a satellite can monitor an area roughly 60°N, S, E and W of the sub-satellite point directly below. A small number of geostationary Meteosat satellites, for example Meteosat 7–11, can ensure global coverage. Imagery is sensed by a nodding mirror using satellite spin to scan a series of E–W lines, each transmitted as it is received. Full imagery is updated every 5–15 minutes, giving near real-time monitoring (figure 9.5).

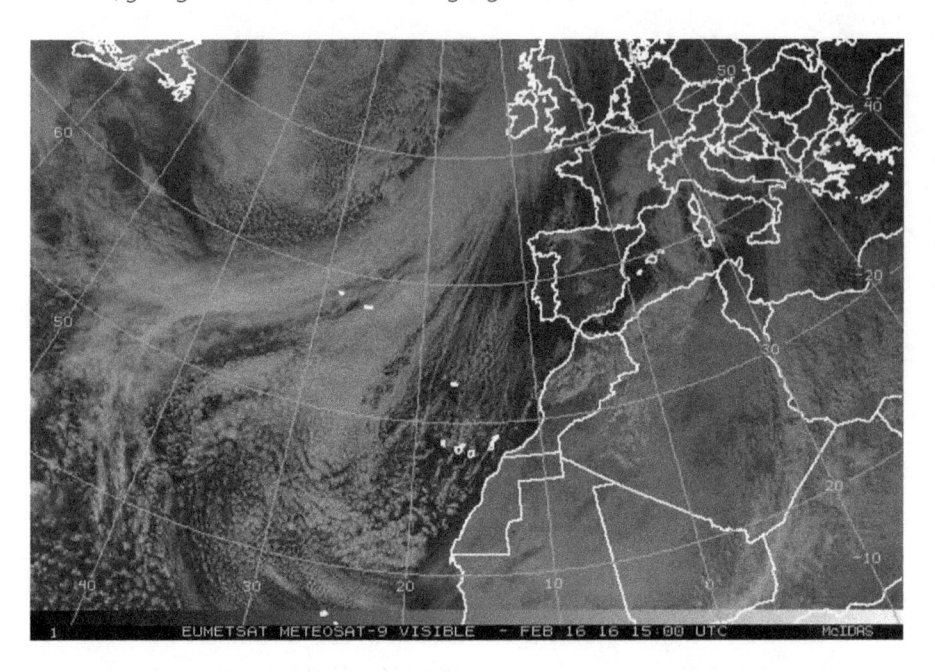

Figure 9.5: *Meteosat image, © 2016 EUMETSAT.*

The GOES series comprises four generations, with the latest satellite, GOES 17, starting deployment in 2018. Currently, GOES 13–17 provide global coverage, albeit with appreciable time of flight delays.

9.6 Comparison of polar orbiting and geostationary satellites

9.6.1 Advantages of polar orbiting satellites

These satellites fly in low orbits, and spatial resolution is good (spy systems can see below 1m resolution). Polar orbiters provide information for all latitudes, especially high ones where scans partly overlap. However, as with **all** electro-optical systems, night applications are limited, and in cloud optical surface imaging is not possible.

9.6.2 Disadvantages of polar orbiting satellites

A satellite is only visible when it is above the radio horizon, so the high Arctic and Antarctic have relatively limited coverage. There is also an interval between successive images of any region on the ground, even with clear skies. They have a relatively narrow FOV of the terrestrial surface while ensuring good resolution of regions within it.

9.6.3 Advantages of geostationary satellites

These satellites provide images of the same region regularly over a wide FOV. Successive images allow the British Meteorological Office to model short-term weather predictions to high accuracy, and provide longer-term climate-related predictions. Regular updated images are a vital aid in uninhabited or rarely visited areas, such as the South Atlantic. Cloud movements are monitored to obtain wind speed and direction. Satellite imagery has revolutionised ocean monitoring, but won't completely replace aircraft and balloon measurements.

9.6.4 Disadvantages of geostationary satellites

Geostationary satellite detail is poor in contrast with NEO satellites (200–400km above Earth). Earth curvature distorts pictures radially, so above 55°N (which includes Britain) images are of limited use, and give no useful coverage over the Arctic or Antarctic.

9.7 Weather sensing satellites

There are many weather satellites but only two series are selected briefly here to provide context. A list of current and planned satellites is found elsewhere [9.1].

Meteosat series This geostationary programme operated by EUMETSAT acquires visible and reflected infrared images of Europe and Africa. Meteosat 1–7 provides reliable and continuous meteorological observations from space to a large user community. In addition to Earth imagery provision and its atmosphere every 30 minutes in three spectral bands (visible, infrared and water vapour) via the Meteosat Visible and InfraRed Imager (MVIRI) instrument, there are various post-satellite processed meteorological products. Satellites are 2.1m in diameter and 3.2m long. Satellites have an in-orbit mass of 282kg, spinning at 100rpm around their main axis. Meteosat Second Generation (MSG) provides improved user requirements to improve weather forecasting and numerical weather predictions.

They are 3.2m in diameter and 2.4m high, spinning 100rpm anticlockwise at an altitude of 36,0000km.

Radarsat series This is a Canadian unmanned Earth-orbiting satellite that acquires C-band (4–8GHz) radar images. The Canadian Space Agency has promoted the peaceful development of space application of Earth observation to provide regional and global Earth anthropogenic solutions. Canadian satellites RADARSAT 1–2 were used for several years to provide information on maritime surveillance, disaster management and ecosystem monitoring. Currently, Canada is developing its RADARSAT Constellation with several smaller satellites to maximise their ability to conduct 24/7/365 space-based surveillance.

9.8 Earth observation satellites

There are many Earth observation satellites, with a few series selected here. A full list of current and planned satellites is found elsewhere [9.1]. Space-based Earth observation is expensive. The cost (without subsidy) of developing a remote sensing satellite and putting it into LEO is typically £100 million, with the cheapest US$/kg provided by the Falcon Heavy launcher (US$2200/kg).

Landsat series This was the first satellite designed specifically for observing Earth's surface and maritime environment. It was a great success, producing images for marine, agriculture and other applications. Many Landsat satellites have operated since 1972. Landsat 1–3 carried two main sensors: a video-imaging instrument, the Return Beam Vidicon (RBV), and a MultiSpectral Scanner (MSS), in near-polar and sun-synchronous orbits at an altitude of 913km with a repeat cycle of 18 days.

Landsat 4–5 were placed in near-polar, sun-synchronous orbits at a lower 705km altitude, with a 16-day repeat cycle. These carried MSS and another multispectral scanner, the Thematic Mapper (TM), with sun-synchronous periods of 99 minutes, and 98.2° inclination.

Key Landsat 4–5 detection band characteristics are given:

MSS Channel Waveband (Microns) Spatial resolution

1 0.5–0.6 80m; **2** 0.6–0.7 80m; **3** 0.7–0.8 80m; **4** 0.8–1.1 80m

TM Channel Waveband (Microns) Spatial resolution

1 0.45–0.52 30m; **2** 0.52–0.60 30m; **3** 0.63–0.69 30m; **4** 0.75–0.90 30m;

5 1.55–1.75 30m; **6** 10.40–12.5 120m; **7** 2.08–2.35 30m

SPOT series This series is operated by the French CNES. Satellites are placed in near-polar, sun-synchronous orbits at 832km altitude. Each has two High Resolution Visible (HRV) sensors on board, which work on a 'push-broom' system whose scanners have a large array of detectors collecting data as the satellite moves forward, like a broom sweeping a floor. Instruments view the surface immediately below (nadir view), or are directed so they view to one side, up to 27° off nadir. This is important as it means two views of the same area are collected in a short time of each other, by adjacent over-passes. A place viewed directly overhead on one pass can be viewed from the side on the satellite's next pass. As the system takes two images of the same area with different angles, we can create stereoscopic (3D) images. As each eye views the same scene from a slightly different position, the brain creates a 3D picture. Computers create Digital Elevation Models (DEM) from stereoscopic images, so terrain elevation is measured. Images with two different spatial resolutions are produced from SPOT data. The first is a panchromatic (black-and-white) image with 10m spatial resolution. The second is an image from the SPOT multispectral scanner (SPOT XS) which collects three bands, with 20m resolution. SPOT is used for producing topographic maps.

SPOT 1 and 2 sensors are in near polar sun-synchronous orbits with 832km altitude, and 98.7° inclination, 26-day repeat cycles, and 1900kg launch mass.

SPOT-1–2 (HRV) Panchromatic Mode Channel (Microns) Spatial resolution

1 0.51–0.73 10m

SPOT-1–2 (HRV) Multispectral Mode Channel (Microns) Spatial resolution

1 0.50–0.59 20m; **2** 0.61–0.68 20m; **3** 0.79–0.89 20m

Envisat This was a major environmental satellite mission, but is no longer active, although it still orbits in space. The platform carried many sensors operated by different space agencies. It was larger than most EO satellites with a launch mass of 8,000kg, compared with 2,400kg for ERS-2. The satellite platform had two parts:

1. A service module with instruments needed to operate the satellite, and

2. A payload module containing the various sensors for Earth observation.

Envisat provided information about the atmosphere, biosphere, cryosphere (ice), hydrosphere and lithosphere. Payload included two radar instruments, three spectrometers, two radiometers, the first high-resolution space-based interferometer for long-term observation and two instruments for measuring distance. One sensor, GOMOS, monitored atmospheric ozone in the troposphere to the mesosphere, and other chemical species such as bromine oxide. Bromine is a more aggressive chemical species than chlorine. Envisat sits in a sun-synchronous near polar orbit at an altitude of 800km and 98.55° inclination.

Envisat carried several systems: Advanced Synthetic Aperture Radar (ASAR), Global Ozone Monitoring by Occultation of Stars (GOMOS), Laser Retro-Reflector (LR), MEdium Resolution Imaging Spectrometer (MERIS), Michelson Interferometer for Passive Atmospheric Sounding (MIPAS), Microwave Radiometer (MWR), Radar Altimeter 2 (RA2), Advanced Along Track Scanning Radiometer (AATSR), Doppler Orbitography and Radio positioning Integrated by Satellite (DORIS) and SCanning Imaging Absorption spectroMeter for Atmospheric CHartographY (SCIAMACHY).

EOS series Three platforms – Terra, Aura and Aqua – provide global data on the atmosphere, land and oceans. Aqua is studying precipitation, evaporation and water cycle. Aqua, from the Latin word for water, carries six instruments for water studies of Earth's surface and the atmosphere: Advanced Microwave Scanning Radiometer-EOS (AMSR-E) measures cloud properties, SST, near-surface wind speed, radiative energy flux, surface water, ice and snow; MODerate Resolution Imaging Spectroradiometer (MODIS) measures cloud properties, radiative energy flux and aerosol properties, land cover and land use change, fires and volcanoes. This instrument is also aboard Terra; Advanced Microwave Sounding Unit (AMSU-A) measures atmospheric temperature and humidity; Atmospheric InfraRed Sounder (AIRS) measures atmospheric temperature and humidity, land and sea surface temperatures; Humidity Sounder for Brazil (HSB) has VHF band equipment measuring atmospheric humidity; and Clouds and the Earth's Radiant Energy System (CERES) measures broadband radiative flux. Aqua had 2,850kg mass at launch and deployed with solar panels unfurled is 4.81 × 16.70 × 8.04m.

Jason-1–2 This is a satellite oceanography mission to monitor global ocean circulation and study oceans and the atmosphere, improving global climate forecasts and prediction and monitoring events such as El Niño and ocean eddies. Jason-1 has five instruments: Poseidon 2 is a nadir-pointing radar altimeter using C and Ku bands for measuring height above sea surface. The Jason Microwave Radiometer (JMR) measures water vapour along altimeter paths to correct for pulse delay. DORIS determines orbit within 10cm, and provides ionospheric correction data for Poseidon 2.

TOPEX TOPEX/Poseidon was a joint satellite mission between NASA and CNES to map ocean surface topography. The first major oceanographic satellite in space, TOPEX/Poseidon revolutionised oceanography, proving the value of satellite ocean observations. A malfunction ended normal operations in 2006, although past data archives are still available.

TerraSAR-X This is an active radar sensor providing information about ships and sea ice. TerraSAR can work with other Airbus Defence and Space satellites, such as SPOT 6–7 and Pleiades 1a–1b. TerraSAR-X offers a Wide ScanSAR mode with large area coverage (270 × 200km) and 40m moderate spatial resolution. Delivery time of non-quality controlled ship detection can be as little as 20 minutes after acquisition. In addition to good temporal sampling, it provides daily revisits. Oil spillage detection is now an important global concern (such as the Deepwater Horizon oil spill in the Gulf of Mexico in 2010). A pan-European surveillance system for European waters includes EMSA, Frontex and Community Fisheries Control Agency (CFCA) to detect oil pollution. Sea ice and iceberg detection has been of maritime concern for many years and past data is available from several medium spatial resolution SAR satellites, such as Envisat and Sentinel-1, but TerraSAR-X provides greater discrimination between smooth and rough ice and identification of the best routes for vessels.

International Space Station This isn't strictly a satellite, but a manned platform taking Earth observation images. It possesses a dedicated EO platform for data acquisition, and imaging photography with the Cupola observation module. Its seven windows are used to conduct experiments, dockings and Earth observations. It was launched aboard Space Shuttle mission STS-130 in 2010 attached to the Tranquility (Node 3) module.

Sentinel programme ESA is developing seven missions under the Sentinel programme, including radar and superspectral imaging for land, ocean and atmospheric monitoring. Sentinel-1 provides all-weather, day and night radar imaging for land and ocean services. Sentinel-2 provides high-resolution optical imaging for land services (such as imagery of vegetation, soil and water cover, inland waterways and coastal areas) and monitors information for emergency services. Sentinel-3 provides ocean and global land monitoring services. Sentinel-4 is embarked as a payload upon a Meteosat Third Generation Satellite, providing data for atmospheric composition monitoring. The first will launch in 2021. Sentinel-5 Precursor was launched successfully in October 2017. Its primary purpose is to reduce the data gap (especially SCIAMACHY atmospheric observations) between Envisat in 2012 and the launch of Sentinel-5 in 2021. Sentinel-5 will provide data for atmospheric composition monitoring. Sentinel-6 is intended to sustain high-precision altimetry missions following the Jason-2 satellite.

9.9 High-resolution satellite missions

Over the next few years, more high-spatial resolution systems and hyperspectral sensors will launch, some providing real-time video applications. Several commercial consortia are developing missions, including Earthwatch, Space Imaging, EOSAT and Orbimage. 'Consortia' is a word describing a group of organisations that get together for a particular business scheme. Large companies such as Mitsubishi, Kodak, Lockheed and Hitachi are major partners within these consortia. Until now, satellites run by commercial organisations have not provided images with the same detail as aerial photography. The aerial photography market is estimated to be $2 billion per year, while the satellite imagery market is currently estimated at around $500 million per year (NRSC Ltd). High-resolution satellite imagery provides an alternative to aerial photography, with applications benefiting from high-resolution data.

9.10 Small satellite missions and nanosats

Small satellites or nano-satellites weigh much less than bigger missions such as Meteosat. There are many advantages in using smaller, lighter satellites. They are cheaper to build, launch and operate. They carry a smaller payload and are launched for specific purposes. One of the world leaders in this technology is Surrey Satellites, founded by Sir Martin Sweeting, and its subsidiary DMC International Imaging (DMCII), providing disaster data to users worldwide for emergency mitigation. Small satellites are generally launched by 'piggy-backing' as spare capacity on other larger satellite launches. The smaller the mass of an individual satellite, the more small satellites that can be launched with a single large satellite. Small-sat categories in terms of mass, and power requirements are given in table 9.3:

Category	Mass	Power (W)
Nanosatellite	<10	<10
Microsatellite	10–50	100
Minisatellite	50–500	500
Small satellite	500–1,000	1,000
Large satellite	>1,000	>1,000

Table 9.3: *Small satellite missions.*

9.11 Other notable Earth observation satellites

There are many Earth observation satellites: a few are listed briefly here by country and space agency.

Brazil Amazônia (land resources) and CBERS (China/Brazil Earth Resources Satellite) series providing global coverage.

Canada UrtheCast: UrtheCast 1–2, OptiSAR (Radar 1–8 and Optical 1–8).

China Gaofen series (optical satellites with 2m panchromatic resolution), TanSat (CarbonSat carrying a high-resolution CO_2 spectrometer) and Yaogan series (government remote sensing satellites with likely military reconnaissance applications).

European Space Agency (ESA) Aeolus, for wind speeds and wind profiling, with active Doppler wind lidar.

ASIM (Atmosphere-Space Interactions Monitor) to study transient luminous events and terrestrial gamma ray bursts).

CryoSat 1–2 uses radar altimeters to determine continental ice sheets and marine sea ice cover thickness with three instruments: the SIRAL altimeter, DORIS receiver and Laser reflector.

EarthCARE (Earth Clouds, Aerosols and Radiation Explorer), a mission to understand the interactions between cloud, radiative and aerosol processes regulating climate.

SMOS (Soil Moisture and Ocean Salinity) monitors vegetation water content, snow cover and ice structure.

France CALIPSO (Cloud-Aerosol Lidar and Infrared Pathfinder Satellite Observations) features a two-wavelength (532 and 1064nm) polarisation-sensitive lidar to measure aerosol and cloud properties.

Jason series, for high-precision oceanographic altimetry.

Pléiades series, high-resolution optical imaging.

Germany GRACE (Gravity Recovery And Climate Experiment) composed of two identical satellites flying 220km apart in a 500km polar orbit.

TerraSAR-X series, 1m radar resolution class with image, stripMap 3m, and ScanSAR 16m resolution modes.

India IRS series, the first true Indian state-of-the-art remote sensing satellites.

ISRO: Cartosat series, high-resolution panchromatic, multispectral and hyperspectral Earth observation.

Oceansat series, providing ocean colour data with global wind vector, and characterisation of the lower atmosphere and ionosphere.

RISAT (Radar Imaging Satellites) series, imaging with active X-band SAR.

Italy COSMO 1–4, with COSMO-SkyMed Second Generation providing dual-use (civilian and military) high-resolution imagery globally and the Mediterranean particularly.

Japan ADEOS series (ADvanced Earth Observing Satellite), includes an Advanced Visible Near Infrared radiometer (AVNIr) and Ocean Colour and Temperature Scanner (OCTS), and six further sensors including TOMS (Total Ozone Mapping Spectrometer) and ILAS (Improved Limb Atmospheric Spectrometer), an improved spectrometer for measuring infrared radiation on the edge of the atmosphere.

TRMM (Tropical Rainfall Measuring Mission) carries four instruments, including the TRMM Microwave Imager (TMI) and Lightning Imaging Sensor (LIS).

National Aeronautics and Space Administration (NASA) Aquarius, in conjunction with the Space Agency of Argentina (to measure global sea surface salinity).

CIRAS (CubeSat InfraRed Atmospheric Sounder), to measure upwelling IR radiation from Earth in the MWIR region.

CIRiS (Compact Infrared Radiometer in Space), a 7–13 micron IR imaging radiometer.

CLARREO series (CLimate Absolute Radiance and REfractivity Observatory).

CloudSat, uses mm wavelength radar to measure cloud altitude and properties.

GEDI (Global Ecosystem Dynamics Investigation), an ISS attached instrument providing the first high-resolution laser ranging observations of forest canopy height and canopy height structure.

ICESat 2 (Ice, Cloud and land Elevation Satellite), designed to measure polar ice sheets.

OmniEarth 1–18, a planned 18-satellite constellation providing high-resolution multispectral imagery to commercial and governmental customers.

ORBIMAGE: OrbView series, GeoEye's next generation of 0.41m resolution panchromatic and 1.64m multispectral colour imagery.

Planet Labs: Dove and Flock series with dozens of satellites to provide 3–5m resolution images; RapidEye 1–5 for commercial Earth imaging 6.5m resolution.

Skybox Imaging: SkySat series for high-resolution panchromatic and multispectral Earth imagery.

Space Imaging: IKONOS 2 retired in 2015 after 15 years' service, providing 1m ground resolution, became part of GeoEye, and then DigitalGlobe group.

Spire Global Inc. Lemur series, meteorology and ship traffic tracking payloads.

Russia Almaz series (radar providing 5–7m resolution and imagery resolution (2.5–4m) and Zenit series, now retired (carrying reconnaissance cameras).

UK Earth Observation (UKSA was BNSC): UK-DMC Disaster Monitoring Constellation, imaging several thousand km² with 22–32m ground sampling distance.

Earth-I: Vivid-i 1–5, a future five-satellite constellation for sub-metric full-colour video/still imaging.

NovaSAR-S: small SAR missions for flood monitoring, agriculture, forest monitoring, land use mapping and disaster management, as well as maritime applications.

SSTL: Carbonite 1–2, 1m high-resolution EO satellite with 1m panchromatic and sub-4m multispectral modes.

University of Surrey: UoSat series, includes various multispectral cameras and high-resolution panchromatic cameras.

USA BlackSky Global: BlackSky series (1m resolution microsatellites).

Capella Space: Capella 1, first of a 30-satellite SAR constellation to provide hourly imagery.

DigitalGlobe (EarthWatch): WorldView series (the world's only 0.5m resolution satellites), GeoEye series, and QuickBird (1 and 2, decayed in 2015) and EarlyBird satellites.

Harris: HyperCube and HySpec series, providing wind speed, direction and elevation.

Questions

9.1 As a function of height above Earth's surface gt is given by:

$$gt = \frac{g_0 R^2}{(R+H)^2}$$ where R is Earth's radius (6400km) and H is the height above Earth's surface in km. Find the acceleration due to gravity in terms of g_0 at an 1200km orbital height (4 significant figures).

9.2 An object of mass m is in orbit around Earth where the force of gravity equals the centrifugal force F_c given by:

$$F_c = \frac{\text{object mass} \times \text{orbital velocity}^2}{\text{distance to the centre of rotation}^2}$$

If the orbital velocity is related to height so that:

$$\text{orbital velocity} = \sqrt{\left(\frac{g_0 R^2}{R+H}\right)},$$

find the orbital velocity at a height of 1,200km (4 significant figures).

9.3 Compare a polar orbiter with a geostationary satellite. What are the advantages of a polar orbiting satellite?

9.4 Find the minimum circular orbit for Earth, radius = 6,380km.

9.5 Calculate the work done, or equivalent potential energy, by a satellite leaving Earth's surface to travel to infinity.

9.6 To achieve escape velocity, what is the *minimum* speed we must give a satellite?

9.7 By considering potential energy, find an asteroid's speed when it reaches Earth's surface, if it is heading directly towards Earth at 13kms^{-1} relative to the planet when the asteroid is 11 Earth radii from Earth's centre. Neglect atmospheric effects. M_{Earth} = 5.98 × 10^{24}kg, R_{Earth} = 6378km, and G = 6.67 × 10^{-11}m^{3}kg^{-1}s^{-2}.

9.8 In the absence of gravitational and frictional forces:

$\Delta V = U \ln(M_i/(M_i - M_f))$, where U is the exhaust gas velocity w.r.t the rocket. Estimate the fuel percentage required for a satellite to achieve LEO velocity of ~7.5 kms^{-1}, and U ~2.3kms^{-1}.

9.9 Considering Kepler's third law, explain the consequence of doubling the orbital height on the orbital period.

9.10 Consider a nanosatellite with 3kg mass in a circular LEO with velocity = 7774ms^{-1}. Calculate the centripetal force on the satellite, the satellite altitude and the satellite period.

Reference

[9.1] space.skyrocket.de/directories/sat.htm

10
Introduction to Satellite Image Processing and Other Imagery Sources

'From human imagination to actualised Earth Observation, we now image distant worlds and remote places, as familiar as our own homes.' Chris Lavers

10.1 Introduction to image processing

In this chapter, we will examine the fundamentals of Earth observation image processing and sources of imagery/data, augmenting that from satellites. Digital processing is an important aspect of remote sensing, allowing precise identification of surface features and rare mineral deposits (see Chapter 7). Image processing is used for various data: gamma-ray images from the Cherenkov Telescope Array, X-rays from the Chandra X-ray Observatory (CXO), UV, NIR and SAR, as well as optical. Today, most remote sensing data are recorded digitally, so almost all image interpretation and analysis involves digital processing aspects, data correction procedures, and enhancement to facilitate interpretation and classification. Typical image processing functions include: Pre-processing, Enhancement, Transformation and Classification Interpretation and Analysis.

10.2 Pre-processing

This requires operations before the main data analysis and information extraction, such as radiometric and geometric corrections. Radiometric corrections include correcting data for sensor irregularities and unwanted noise, and converting data so they accurately represent reflected or emitted radiation measured by sensors, introduced in section 5.10. Geometric correction includes variations in sensor-Earth geometries, and data conversion to standardised co-ordinates on Earth's surface, usually done by the supplier.

All remote sensing imagery is inherently subject to geometric distortions caused by factors including scanner perspective motion, platform altitude, attitude, velocity, terrain relief and curvature and Earth rotation. Detector response varies depending on the specific sensor and platform and on acquisition conditions. Variation in geometry and illumination between images may be corrected by modelling the geometric relationship and distance between Earth, sun and sensor, section 5.10 [5.3]. Corrections ensure the geometric image representation is accurate. Some predictable variations are easily modelled; however, unsystematic or random errors cannot be modelled and corrected in this way. *Geometric registration* involves identifying co-ordinates of typically four recognisable ground control points in multiple band images and matching them to their true positions and to each other. Co-registration is performed by registering images to a reference image, prior to performing image transformation procedures.

To geometrically correct the original image, resampling takes input Digital Number (DN) values to generate new corrected output digital values. There are three common methods for resampling: bilinear interpolation, nearest neighbour and cubic convolution. Bilinear interpolation takes a weighted average of four pixels in the original image nearest the new pixel location. Averaging nearest neighbour values alters the original pixel values and creates new digital output values. This is unwanted if further processing and classification based on spectral response is done, in which case resampling is best achieved after classification (post-processing). Cubic convolution is more complex and calculates a distance weighted average of a block of 16 pixels from the original image surrounding the new output pixel location, resulting in completely new pixel values. These methods produce images with a sharper appearance. Nearest neighbour resampling uses the pixel DN value in the original scene nearest the new pixel location in the corrected image, and is the simplest method, as it doesn't alter pixel values, but may result in pixel value duplication, while others are lost. However, both cubic convolution and, to a lesser extent, bilinear interpolation create dark edge cells, by averaging the many zero values located around the perimeter.

Scattering occurs as radiation passes through and interacts with the atmosphere, which reduces or attenuates energy illuminating the surface. In addition, the atmosphere attenuates signals propagating from the target to the satellite or UAV. Various atmospheric correction methods are applied, from detailed modelling of atmospheric conditions during acquisition [5.4–5.5] or calculations based on image data, such as removing scatter by subtracting the minimum observed values in 'dark' areas, assuming a dark object is noise and should be close to zero if the atmosphere is clear, such as 'dark' returns from water (figure 10.1 bottom – see plate section). However, noise can occupy high pixel levels (figure 10.1 top – see plate section), while real features such as the housing areas of Iceland's capital Reykjavik appear to be 'noise-like' in appearance (figure 10.1, coastal region bottom – see plate section).

Noise includes systematic banding or dropped lines, which must be corrected before further enhancement or classification is performed. Striping was common in early Landsat MSS data due to drift and variations in the response of the six MSS detectors. Drift was different for each detector, causing the same brightness to record differently. A relative correction among the sensors brings their apparent values into agreement with each other. Dropped lines are corrected by replacing the faulty line with the pixel values in the line above or below, or the average of them. Consider a small data region (8 × 8 pixels) over water from the bottom half of figure 10.1, shown in figure 10.2. Data is recorded using 9 bits with 0–511 values (512 levels). If noise in the dark region has a value of 50, subtraction of 50 from each pixel reduces the impact of noise.

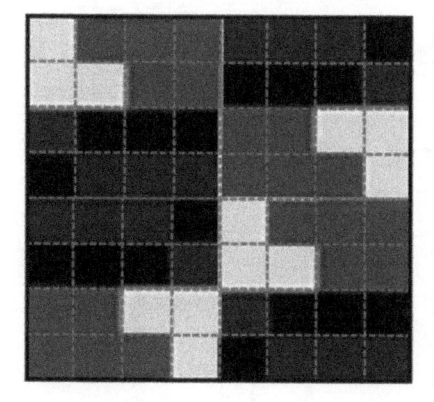

490	200	200	200	50	50	50	50
490	490	200	200	50	50	50	50
50	50	50	50	200	200	490	490
50	50	50	50	200	200	200	490
50	50	50	50	490	200	200	200
50	50	50	50	490	490	200	200
200	200	490	490	50	50	50	50
200	200	200	490	50	50	50	50

Figure 10.2: *Small pixel region values over water.*

10.3 Image enhancement

Image enhancement improves appearance to assist interpretation and analysis. Examples include contrast stretching to increase tonal distinction between features and spatial filtering to enhance or reduce spatial patterns. Enhancement will manipulate the image DN pixel values, even if radiometric corrections for illumination, atmospheric influences and sensor characteristics are done prior to data distribution to users. With large spectral response variations from real surfaces, such as grasslands, deserts, ice, water etc, no generic radiometric correction can account for optimum brightness range for **all** surfaces. Thus, each image requires an adjustment of range and brightness values. In raw imagery, useful data often occupy only a small part of the available range of digital values. Contrast enhancement involves changing the original values so more of the available range is used, increasing contrast between surfaces. Consider an image histogram, figure 10.3a. A histogram is a graphical representation of image brightness values. Brightness values 0–511 are displayed along the x-axis. The frequency of values is shown on the y-axis.

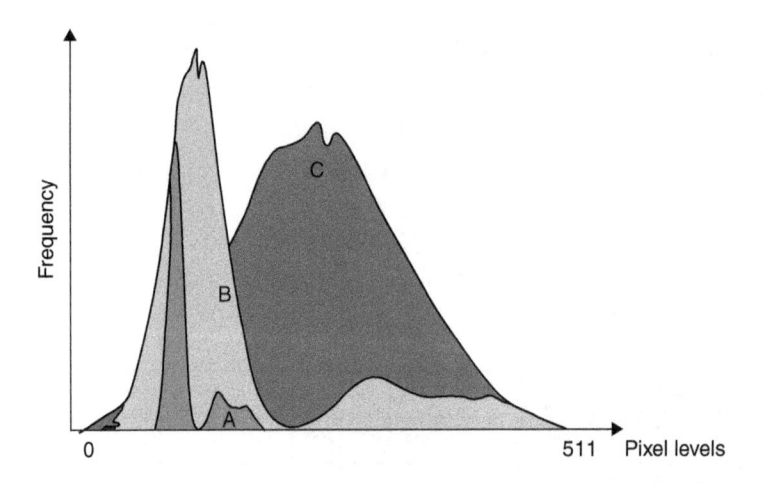

Figure 10.3: (a) *Histogram range levels occupied (100–200)*, (b) *linear contrast stretch of these levels to all (0–511) levels, and* (c) *histogram-equalised stretch of forest pixel occupied levels (100–135) to all (0–511) levels.*

By manipulating the value range, graphically represented by its histogram, we can enhance data. The simplest enhancement is a *linear contrast stretch*, which usually uses the minimum and maximum brightness values, lower and upper bounds, and applies a transformation to stretch the occupied range to the full range – for example, the minimum value occupied by actual data in the histogram is 100 and the maximum value 200. These 100 levels occupy less than one-fifth of the 512 levels available. A linear stretch uniformly expands this small range to cover the full range of values from 0 to 511. This enhances contrast in the image, with light-toned areas appearing lighter and dark areas appearing darker. A linear contrast stretch is usually required just to see detail even on a bright computer display image, as reflectance levels from the Earth are often very dim.

A uniform distribution of input values across the full range isn't always appropriate. In this case, a *histogram equalised stretch* may be better. For example, suppose we have an image covering part of a forest, and the forest occupies the digital values from 100–135 (first peak of A). If we wish to enhance the forest, we stretch only that part of the histogram represented by forest 100–135, to the full greyscale range 0–511. All pixels below or above these values are assigned 0 and 511, and detail in such areas is lost. However, forest detail is enhanced.

10.3.1 Spatial filtering

This encompasses digital processing functions that highlight or suppress specific features based on spatial frequency. Filtering may involve moving a window a few pixels wide in dimensions over each image pixel, applying a mathematical calculation using the pixel values under the window and replacing the centre pixel with the new value. The window is moved in row and column dimensions, one pixel at a time, and the calculation repeated until the entire image is filtered and a new image generated. By varying the calculation performed and weightings of individual pixels in the window, filters are designed to enhance or suppress different features.

10.3.2 Matrix filter approach

Point, line and edge feature detection depends critically on contrast in the neighbourhood of visible features. We could define a high-pass filter to initially locate building outline when searching massive area images otherwise reflectance saturated, removing low frequency spatial pattern background detail [10.1–10.2]. Appropriate filters negatively weight neighbour pixel coefficients to enhance regions, so fine detail is emphasised. Edge or boundary extraction is essential to building detection. Filters may sharpen bright regions more than dark ones and flag up areas for further manual scrutiny, with reduced false alarms.

We define a mask weighting a high-pass filter to the local mean. The filter approach could take a 3 × 3 image field window centred on Z_5:
$$\begin{matrix} Z_1 & Z_2 & Z_3 \\ Z_4 & Z_5 & Z_6 \\ Z_7 & Z_8 & Z_9 \end{matrix}$$
(eq 10.1)

For example, with an equal weighted mask the filter is:
$$\begin{matrix} 1 & 1 & 1 \\ 1 & 1 & 1 \\ 1 & 1 & 1 \end{matrix}$$

so: $Z_{average} = 1/9 \sum_{i=1}^{9} Z_i$

The central window pixel is termed the *origin*. For a 3 × 3 sample centred on a window pixel (origin) of say 100: $\begin{matrix} 0 & 50 & 0 \\ 50 & 100 & 50 \\ 0 & 50 & 0 \end{matrix}$ then:

$Z_{average} = 1/9(0+50+0+50+100+50+0+50+0) = 33$

A weighted *sharpening* high-pass filter will perform a sliding neighbour operation on the origin:

$\begin{matrix} -1 & -1 & -1 \\ -1 & 9 & -1 \\ -1 & -1 & -1 \end{matrix}$ **(eq 10.2)**

so that:

$Z_{average} = 1/9(-1´0 - 1´50 - 1´0 - 1´50 - 1´50 + 9´100 - 1´50 - 1´0 - 1´50 - 1´0) = 77.8$.

a higher resultant value for the origin pixel with the high-pass filter than without it. This simple high-pass filter may be considered as a convolution application of spatial filtering performed on the original, unprocessed image.

Application of the high-pass filter above, performing a sliding neighbour operation allows searching massive reflectance saturated imagery areas to locate building outlines and remove low frequency spatial pattern background. The original image is shown in greyscale (figure 10.4a – see plate section). An absolute change plot for the high-pass filter is shown (figure 10.4b right – see plate section) and considered an acceptable way to improve edges as ± built/shadow changes combine, enhancing high-built (straight) edges. This shows greyscale change, eliminating much of the background terrain (homogeneous light and dark regions). A simple high-pass filter, displayed in modular form, shows vegetation (negative dark areas) and most change occurring at the scale mid-point (zero level). Roofs give large ± edge changes, from high (saturated) reflectance, plunging into adjacent dark shadow. Vegetation edges are less dramatic but still have shadow. The filter removes large bright areas (for example, the SW quadrant below the long white buildings) while retaining defined built edges. High-reflectance regions next to low reflectance regions produce a greater uniform area across the image, for example, near (275, 50).

Varied range limits (saturation threshold) discriminate sharp defined edges (such as buildings) and, to a lesser extent, vegetation and tracks (more jagged edges).

No current literature has discussed the effectiveness of such a high-pass spatial filtering approach against other techniques for man-made land clearances or humanitarian applications using high-resolution satellite imagery.

One alternative to manipulations in the spatial domain is analysis of the frequency domain. With a frequency domain approach, images may be separated into frequency components with a suitable Fourier transform operation. A recent comparative study of image enhancement techniques related to general image fusion [10.3] proposed evaluation metrics for spatial and frequency methods, including error analysis, but performed no comparative evaluation on imagery.

A low pass filter, such as

$$\frac{1}{9}\begin{matrix} 1 & 1 & 1 \\ 1 & 1 & 1 \\ 1 & 1 & 1 \end{matrix} \qquad \text{(eq 10.3)}$$

emphasises larger, homogenous areas of similar tone and reduces fine detail, smoothing appearance. Average and median filters are often used for radar imagery. High-pass filters do the opposite, sharpening the appearance of fine detail. One implementation of a high-pass filter first applies a low-pass filter and then subtracts the result from the original, leaving the high-spatial frequency information. Direction or edge detection filters highlight linear features, such as roads or field boundaries. Filters are designed to enhance features oriented in specific directions, that is, vertical or horizontal, and are useful in applications such as geology for detection of linear fault trends. Further masks such as sobel operators for horizontal and vertical and more sophisticated sets of masks for all possible edge orientations (Robinson edge detectors), dilation (to fill in gaps) and erosion (to break connections) are used. In fact, a low-pass filter: $\begin{matrix} 1 & 1 & 1 \\ 1 & 1 & 1 \\ 1 & 1 & 1 \end{matrix}$ is useful to apply when line dropouts occur. However, there are different methods that can be best used to rectify IKONOS imagery [10.4].

Another example, in figure 10.5a satellite imagery, shows an area in Ethiopia displayed in NIR with recently built huts to house dam-building workmen. Figure 10.5b shows the same area using the high-pass filter of equation (10.2). The absolute (modular) change of the high-pass mask filter is shown in greyscale in

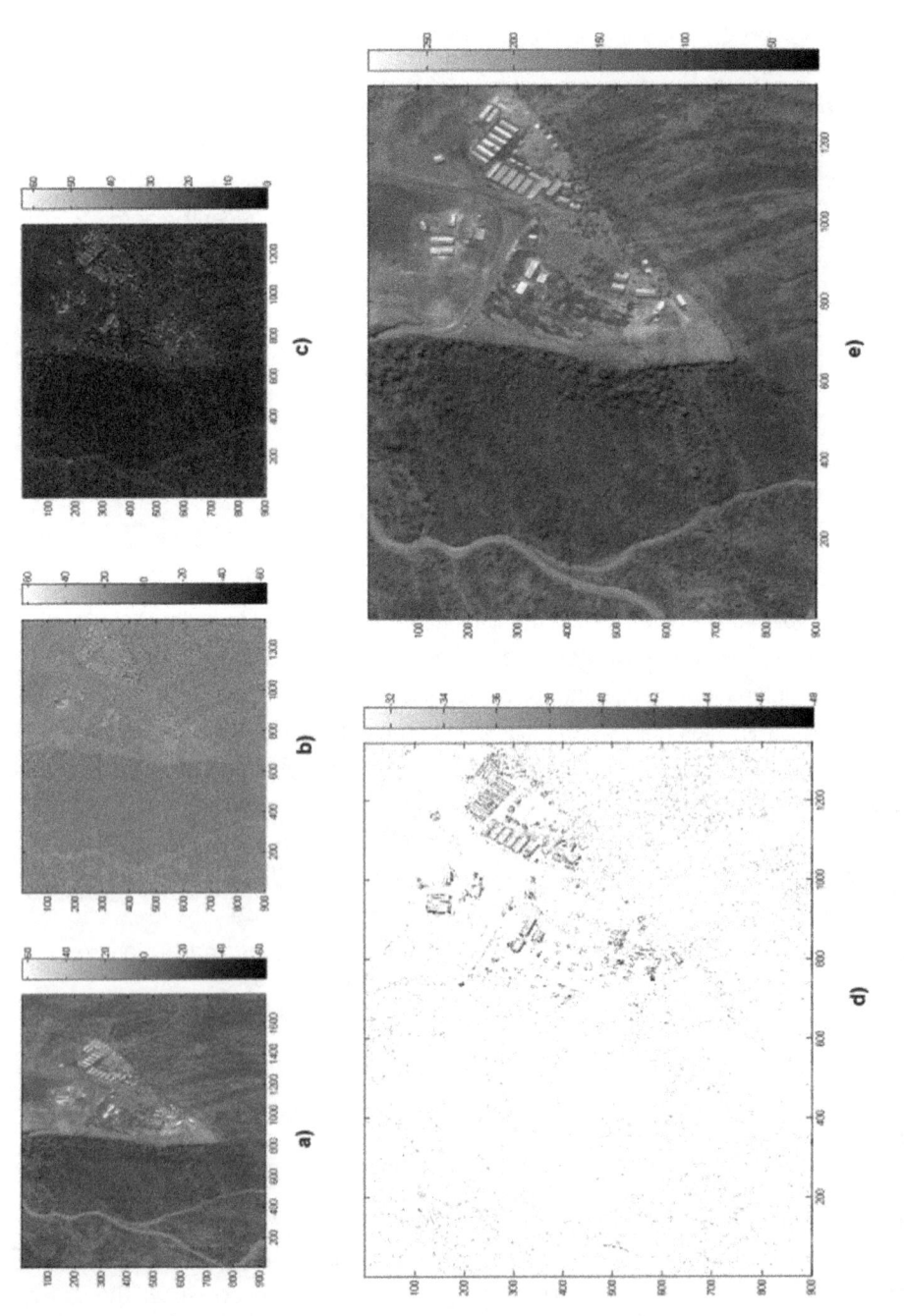

Figure 10.5: *Various low-pass and high-pass matrix filter operations.*

figure 10.5c, but is easier to view in this book if the mask filter 'change' is inverted to show as a little dark on a white background, rather than a little light on a dark background as in the larger figure 10.5d. Figure 10.e shows the application of a 'smoothing' low-pass filter similar to that in equation (10.3).

10.4 Image transformations

Image transformations are usually applied to multiple spectral bands as arithmetic operations – for example, addition or multiplication of bands, transforming the original bands into new combinations that may better display features in applications such as mineral detection. Image transformations involve multiple band manipulation or multispectral images of the same area acquired at different times (multi-temporal). Image transformations generate new images to highlight surfaces of interest better than original ones, and can be termed under *change detection*. Image subtraction can identify changes between images collected on different dates – for example, two geometrically co-registered with pixel brightness values in one image, subtracted from pixel values in another, producing a suitable difference image. Areas where there is little or no change between the original images have resultant brightness values near the mid-scale point, while those with significant change have values much higher or lower, brighter or darker (near 511 or 0 respectively), depending on the reflectance change between images. These types of image transformations are useful for mapping areas where deforestation or land clearance occurs. Image transformations such as reflection

$$\begin{matrix} -1 & 0 & 0 \\ 0 & 1 & 0 \\ 0 & 0 & 1 \end{matrix} \quad \text{(eq 10.4)}$$

and rotation

$$\begin{matrix} \cos\theta & \sin\theta & 0 \\ -\sin\theta & \cos\theta & 0 \\ 0 & 0 & 1 \end{matrix} \quad \text{(eq 10.5)}$$

can also be used very effectively using matrices.

10.4.1 Land clearance case study

High-resolution satellite imagery permits verification of land clearance violations across international borders as a result of unstable regimes or socio-economic upheaval. Without direct access to these areas to validate allegations of human rights abuse, use of remote sensing tools, techniques and data is extremely important. Software-based imagery assessment can potentially quantify radiometrically calibrated NDVI and temporal land clearance changes.

The necessary methodology, seen in figure 10.6, yields the reflectance of pixels at Earth's surface. IKONOS imagery from the GeoEye Foundation archive provided geometrically rectified data [10.5]. An appropriate radiometric calibration was applied to convert the DN values to at sensor spectral radiance, and then converted spectral radiance to at sensor apparent reflectance [10.6]. This correction was followed by the removal of atmospheric effects due to absorption and scattering with the Second Simulation of a Satellite Signal in the Solar Spectrum-Vector (6SV) method used [10.6–10.7], yielding pixel reflectance at the Earth's surface. The 6SV code enables accurate simulations of satellite observations, accounting for elevated targets, anisotropic and Lambertian surfaces, and calculation of gaseous absorption. The code is based on the method of successive orders of scatterings approximations.

This workflow was applied to satellite imagery discussed in a recent journal paper [10.8]. Satellite imagery is seen in figure 10.7 (see plate section), before and after the Porta Farm Zimbabwe land clearance.

Temporal change uses the equation: $C_T = (\rho_{After} - \rho_{Before}) / (\rho_{After} + \rho_{Before})$ **(eq 10.6)** where ρ is the satellite camera's Digital Number. Human rights workers usually work with aerial or ground-based visible single band imagery. Media outlets don't require NDVI. Historically, most military aerial surveillance operations used single image comparison. However, a better choice would subtract before NDVI from after NDVI where available. Temporal comparison highlights residential areas surrounded by subsistence farmland (north) of varied size and shape (figure 10.7c).

Non-radiometric NDVI was also analysed before (figure 10.8a, left – see plate section) and after (figure 10.8b, middle) for at-aperture apparent reflectance values in overlapped regions with IKONOS red and NIR bands, both co-registered in sharp definition using the NDVI workflow diagram and NDVI equation: NDVI = $(\rho_{NIR} - \rho_{Red}) / (\rho_{NIR} + \rho_{Red})$, **(eq 10.7)**, analogous to Landsat's NDVI.

This work considers space-based monitoring for human rights observation and is not concerned specifically with calibration. However, appropriate radiometric calibration (6SV) methods correct at-aperture spectral radiance to provide

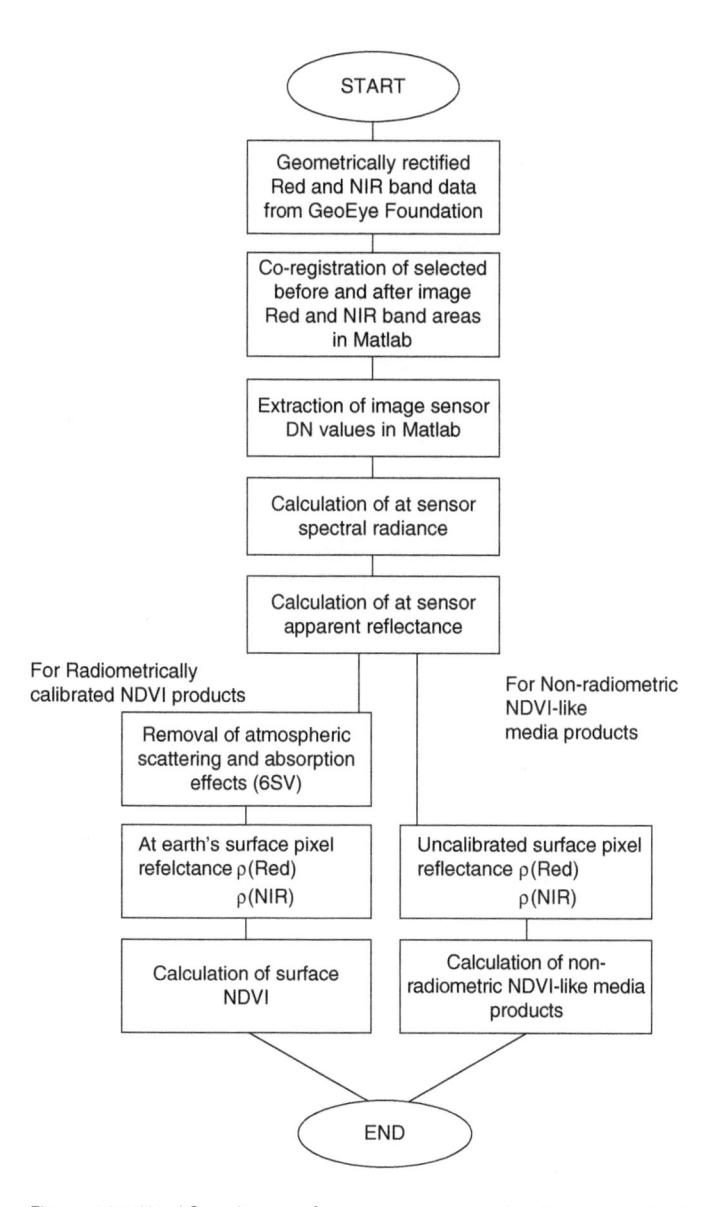

Figure 10.6: *Workflow diagram for co-registration and radiometric calibration.*

normalised NDVI for pixels at Earth's surface for our IKONOS-2 data (figure 10.8c, right). Normalisation changes contrast magnitude but retain features observed previously (see figures 10.8b and 10.8c respectively for comparison). Image division or spectral ratioing is one of the commonest transformations applied to imagery, and highlights subtle spectral response variations in varied land cover. By ratioing data from two different spectral bands, the resultant image enhances variation in spectral reflectance curves between the different spectral ranges, otherwise hidden by pixel brightness variations in each band (for example, figure 10.8), and in geological applications (see Chapter 7).

Healthy vegetation reflects strongly in the NIR while absorbing strongly in the visible red. Surfaces such as soil and water have near equal reflectance in the NIR and red. A ratio of Landsat band 7 NIR divided by band 5 red results in ratios greater than 1.0 for vegetation and ratios around 1 for soil and water; thus, discrimination of vegetation from other surface cover types is enhanced. More complex ratios involving sum and differences between spectral bands were developed for monitoring vegetation conditions. NDVI image transformations monitor vegetation conditions on continental and global scales, and were discussed.

Different multispectral data bands often correlate as they contain similar information. For example, Landsat MSS bands 4–5 typically have a similar appearance since reflectances from the same surface cover types are nearly equal. Image transformation techniques based on complete processing of the statistical characteristics of multi-band data sets can reduce this data redundancy and correlation between bands. One such transformation is called *principal components analysis*, which reduces the number of data bands, compressing much of the information in the original bands into fewer ones. New bands resulting from this statistical procedure are called components. This process maximises the amount of information from the original data into the least number of new components, so the TM seven-band data set is transformed so the first three principal components contain over 80 per cent of the information in the original seven bands. Interpretation and analysis of these three remaining bands, combined visually or digitally, is simpler and more efficient than using all seven original bands.

Atmospheric errors are removed through atmospheric correction. Multispectral data-scattering increases inversely with wavelength, with sources of scattering arising from clouds. Radiometric variations arise from varied land cover. Absolute radiometric correction takes into account measured atmospheric conditions as well as sensor gains and offsets, solar irradiance and zenith angle, to calculate reflectance values as they would have been measured on the ground. The goal is to turn the digital brightness values recorded into scaled surface reflectance. Relative correction is often used in two ways, adjusting individual data bands within a single image, based on subtracting dark object values from each band or normalising bands in images on multiple dates relative to a reference image [10.9]. Relative radiometric correction in single image normalisation uses histogram adjustment, while multiple data image normalisation uses regression. Histogram *contrast stretching* improves image contrast by stretching the range of intensity values to span a chosen range.

Some basic 'grey-level' transformation functions are used for enhancing negative contrast. If **L = 2k** where k is the number of bits used to represent each pixel, such as for a negative image, so

s = (L − 1) −r   (**eq 10.8**)

Where s is the pixel value of the output image and r the pixel value, what value will the top pixel level r = 3 be after the contrast enhancement **s = (L − 1) −r** is applied?

s = (4−1)−3 = 0 so the highest (lightest) level appears dark. Many other linear and logarithmic transformations are possible.

10.5 Image interpretation

Interpretation of remote sensed imagery may involve identification of multiple surfaces. Image interpretation undertaken without 'local knowledge' or field survey is harder with Earth observation; it may lack stereoscopic imagery to obtain height, and may view down, not intuitive to humans who look horizontally. Also, different spectral bands don't appear the same as visible colours seen by the eye.

We can extract information provided by specific 'clues' present in image elements: tone, shape, size, pattern, texture, shadow and usual proximity association.

Tone refers to the relative brightness (in black-and-white images) or a surface colour. Good tone variations allow other elements to be better distinguished.

Shape refers to the overall outline of surfaces, and can be a distinctive means of interpretation. Straight edge shapes represent man-made features such as roads or agricultural fields, while woodland edges are generally irregularly shaped.

Size of features is a function of scale, as a shed, without absolute scale, may be mistaken for a barn in a field! It is vital to assess the size of a surface relative to others in the image, as well as absolute size to aid interpretation – for example, if an area has many large buildings it might suggest industrial use, while smaller buildings separated by green areas might indicate residential use.

Pattern refers to the spatial arrangement of visible surfaces – for example, an orderly repetitive pattern, such as a palm oil plantation of evenly spaced trees.

Texture refers to the image spatial frequency tonal variations. Rough-textured tone changes rapidly while smooth surfaces have little tonal variation. Smooth textures arise from smooth surfaces like roads or grasslands. Rough and irregular surfaces, like forest canopies, result in a rough-textured appearance. Texture is an important tool for distinguishing features using radar.

Shadows provide information on the profile and relative height of changing surfaces. However, shadows such as radar layover can mask areas as well.

Proximity association takes into account the common associations between surface features or near surfaces of interest – for example, a popular water-sports area is often associated with harbours, hotels, piers, marinas and beaches.

10.6 Change detection

Change detection is useful in land use change analysis, for assessing deforestation and snow-melt and for disaster monitoring etc [10.10]. Recently spectral mixture analysis, artificial neural networks and integration of GIS with remote sensing data are becoming important for change detection applications [10.11]. Image thresholding, differencing, ratioing, regression, change vector analysis, vegetation index differencing, multi-temporal principal component analysis and post classification are also used. Change vector analysis considers the magnitude and direction of the spectral change vector, which describes the direction and magnitude of change from the first to second date. Vegetation indices have long been used for monitoring vegetation temporal changes [10.12].

Vegetation mapping is done using a spectral library containing reflectance spectra collected from field measurements.

Automated change detection is followed by expert interpretation, along with field measurements and verification. The amount of expert interpretation required is reduced by developing automated methods.

10.7 Image classification and analysis

Classification is usually performed on multi-channel data sets (assigning pixels in an image to particular vegetation classes, based on their brightness), namely by supervised and unsupervised classification. Digital image classification tries to classify individual pixels based on their spectral information, and is termed spectral pattern recognition, assigning all pixels to particular classes – for example, grass, oak, forest etc – so the resulting classified image comprises a mosaic of pixels, providing a thematic colour map of the original image.

We distinguish between information and spectral classes. Information classes are those an Earth observer requires to identify forest or rock types. Spectral classes are groups of pixels with near similar brightness in that spectral channel. Thus, three phases of operation exist: (i) selecting the training area, (ii) computing mean and covariance information and (iii) assigning each pixel to a class with the highest probability, using statistical and non-statistical methods. The objective is to match spectral classes in the data with the information classes. In a supervised classification, the analyst identifies similar homogeneous representative samples of the different surface cover types, which are referred to as training areas. Selection of appropriate training areas is based on the analyst's familiarity with the geographical area and knowledge of surface cover types present. The analyst supervises the categorisation of specific classes. The numerical information in all spectral bands for pixels comprising these areas trains the software to recognise spectrally similar areas using specific algorithms to determine the numerical signature for each training class. Each image pixel is compared with these signatures and labelled as the class it most closely resembles.

With unsupervised classification after *detection* of objects of interest, *segmentation* will extract areas of interest according to set criteria. Spectral classes are grouped first, based on numerical data, and matched by the analyst to information classes. Clustering algorithms determine the grouping of data structures. The analyst specifies how many groups or clusters are to be searched for, and may specify parameters related to the mean vector separation distance among clusters and variations (covariance) within each. The supervisor may need to break down or combine clusters, with further application of clustering algorithms. Unsupervised classification is thus not entirely without human intervention.

The simplest classifier is perhaps one based on minimum distance (Euclidean), where a unique spectral signature across several bands is compared with a candidate pixel. There are also other distance and likelihood methods. In Euclidean space, the probability of a candidate pixel belonging to a group is based on the covariance among the spectral information being as low as possible. For example, find the Euclidean distance for the following candidate pixel against Signature 1 and Signature 2 training pixel values:

Band spectral	Signature 1 DN value	Signature 2 DN value	Candidate pixel
1	12	10	11
2	17	23	21
3	28	26	25

Table 10.1: *Band spectra, signature and candidate pixel values.*

$$ED_{signature\,1} = \sqrt{(12-11)^2 + (17-21)^2 + (28-25)^2} = 5.10$$

$$ED_{signature\,2} = \sqrt{(10-11)^2 + (23-21)^2 + (26-25)^2} = 2.45$$

The minimum distance calculation doesn't use covariance information and can be performed when there are only a few training pixels. However, the likelihood is the pixel belongs to spectral signature 2. If the spectral signature was exactly like the candidate pixel band values, then ED = 0. In reality, a normal variation for a group of pixel values is likely to occur.

10.8 Other imagery sources

Today there are many good sources of data besides satellite and aircraft imagery to assist the Earth observation specialist. Recent professional growth in commercial UAV suppliers and surveyors such as AirGon, Airobot, Atmos UAV, the GeoCue Group and RUAS provide imagery routinely. UAVs carry varied payloads, such as cameras with microphones, night vision, thermal, chemical and meteorological sensors to measure wind, temperature and humidity, besides other sensors, and can operate with many kg payloads. Due to the wide range of countries' and companies' classifications, it is suggested that the term 'large drone' for fixed-wing **and** multirotor drones should be standardised between 25 and 100kg. Anything less should be considered a small drone, including mini-drones, which may weigh only a few grams.

Web camera networks are also of great value to both those viewing from above and those on the ground. Networks can be governmental or private, and include traffic management [10.13], river networks [10.14], port and harbour facilities [10.15], as well as natural parks and sites of tourist significance and combined with weather information [10.16–10.19].

The advent of social media and smartphones provides further imagery sources with geolocation information valuable to first responders in the case of humanitarian disasters or field geologists. Professional man-portable 360° camera scanning and lidar-based systems can 'stitch' seamlessly together the less than optimal man-gathered information, often on foot, from overlapping image swaths into a complete mosaic. Optimal use of these imagery sources, where available, will help extract the maximum amount of useful information.

Consider imagery of the city hall (Radhus Reykjavikur) looking from the north-east corner of Reykjavikurtjom lake, Reykjavik, Iceland. Satellite imagery (figure 10.8a) can be combined with a web-camera view of the site (figure 10.8b) as well as various 'ground-truth' views from cameras or smartphones (figure 10.8c–d). The synergy of this imagery provides greater situational awareness, relevant for civilian and military applications, as does accessibility to public and private CCTV cameras, and smart TVs from which cameras may be accessed remotely. Developments should be of concern to those interested in the compromise of personal privacy and freedom from intrusive government.

10.9 Web-based satellite image sources

Much good web-based software is available for processing satellite imagery, as are satellite software imagery sources mentioned here. However, readers should be aware that websites may not always be accessible, especially as companies merge. A range of imaging processing software exists from various suppliers, requiring regular updates as applications and software developments take place. Software products currently available include those from:

Figure 10.9: *Imagery possible at different times of the year:* (a) *Sentinel-1 imagery over Reykjavik, Iceland, Sentinel-2 data courtesy of Copernicus/ESA, with visualisation by Pixalytics Ltd.;* (b) *Web-camera picture of the lake [10.18]. Ground truth lake images:* (c) *Handheld camera;* (d) *phone camera (social media source).*

Hexagon AB's ERDAS, a global technology group based in Sweden, which markets precision measuring technologies with over 35 different brands worldwide.

ENVI (Environment for Visualising Images), a software application used to process and analyse geospatial imagery. It is commonly used by remote sensing professionals and image analysts. ENVI combines several automated scientific algorithms for image processing [10.20].

ArcGIS is a GIS system for working with maps and geographic information. It is designed for creating and using maps, compiling geographic data, analysing information using maps and geographic information in many applications, and managing geographic information databases (example shown in figure 10.10). The system provides an infrastructure for making maps and geographic information available openly on the web [10.21].

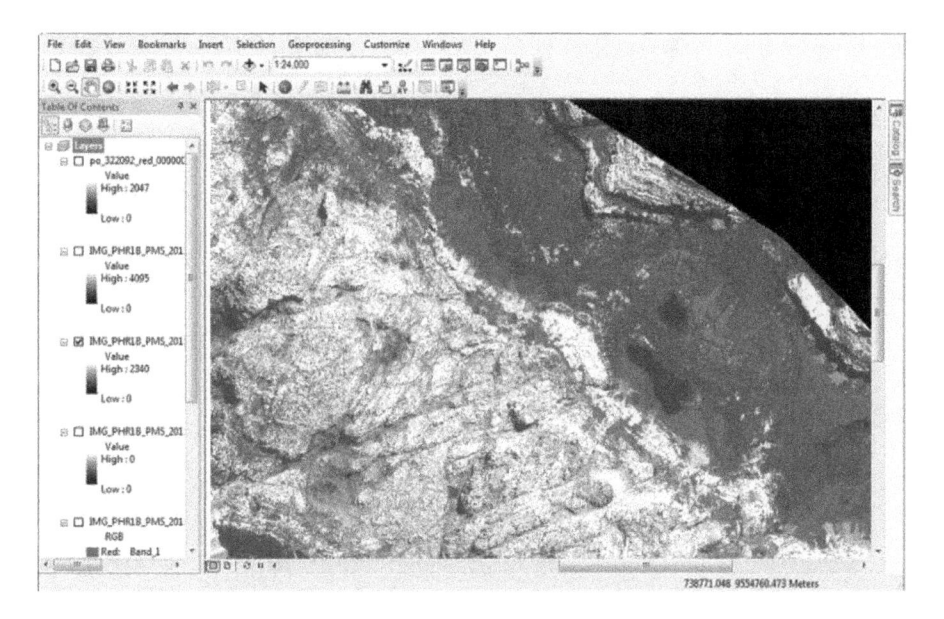

Figure 10.10: *ArcGIS user interface with opened large tiff data file, 1.88GB, over West Papua.*

Quantum GIS (QGIS) is a free and open source cross-platform desktop GIS system application that supports viewing, editing and analysis of geospatial data [10.22].

The UNESCO Bilko Project began in 1987, producing software and eight computer-based modules, distributing copies to 500 plus marine science laboratories and educational establishments and over 3,000 individual users in 90 plus countries worldwide [10.23].

Questions

10.1 Why is geometric co-registration important in the use of multiple band images over the same area?

10.2 Consider applying a weighted customised sharpening high-pass filter
−1 −1 −1
−1   7 −1 performing a sliding neighbour operation on any pixel from figure 10.2
−1 −1 −1
to find the output DN value.

10.3 Explain the difference between a linear contrast stretch and a histogram equalised stretch.

10.4 What is the difference between multiple spectral band data and multi-temporal data?

10.5 How does a hyperspectral system provide more transformation band comparisons than a single band system or a multiband system such as Landsat?

10.6 Consider a greyscale transformation used for negative enhancement. If **L = 2k** where k is the number of bits used to represent each pixel, such as for a negative image, where **s = (L − 1) −r**, and s is the pixel value of the output image and r the pixel value, what value will the pixel level r = 2 be after the contrast enhancement **s = (L − 1) −r** is applied for a 2-bit system?

10.7 Download a copy of BILKO and start processing images of your own!

10.8. Explain the difference between a high-pass and low-pass filter.

References

[10.1] *Computer Processing of Remotely-Sensed Images: An Introduction*, 4th Edition, P Mather and M Koch, p.155 (Wiley-Blackwell, Chichester, 2010, ISBN 9780470742389).

[10.2] *Remote Sensing and Image Interpretation*, Thomas Lillesand, Ralph W Kiefer and Jonathan Chipman (John Wiley & Sons, Inc., New Jersey, 2015, ISBN 1118343289).

[10.3] 'A comparative study of image enhancement techniques', S Khidse and M Nagori, *International Journal of Computer Applications*, Vol. 81(15) (2013), pp.28–32.

[10.4] 'Comparison of different methods to rectify IKONOS imagery without use of sensor viewing geometry', OR Belfiore, C Parente, *American Journal of Remote Sensing*, 2(3), (2014), pp.15–19.

[10.5] 'GeoEye-1 Radiance at Aperture and Planetary Reflectance', NE Podger, WB Colwell and MH Taylor, apollomapping.com/wp-content/user_uploads/2011/09/GeoEye1_Radiance_at_Aperture.pdf

[10.6] '6SV Routine', accessed 21 March 2017 6s.ltdri.org/

[10.7] 'Validation of a vector version of the 6S radiative transfer code for atmospheric correction of satellite data. Part I: Path radiance', SY Kotchenova, EF Vermote, R Matarrese and FJ Klemm, Jr, *Applied Optics*, Vol. 45, No. 26. (2006), pp.6762–6774.

[10.8] 'High-resolution IKONOS satellite imagery for normalized difference vegetative index-related assessment applied to land clearance studies', CR Lavers and T Mason, *Journal of Applied Remote Sensing*, 11(3) 035008 (4 August 2017), doi: 10.1117/1.JRS.11.035008.

[10.9] *Introductory Digital Image Processing: A Remote Sensing Perspective*, J Jensen (Prentice Hall Series in *International Journal of Geographic Information Science*, 25 August 1995, ISBN: 9780132058407).

[10.10] 'Digital change detection techniques using remotely-sensed data', A Singh, *International Journal of Remote Sensing*, Vol. 10(6), November 1988, pp.989–1003, dx.doi.org/10.1080/01431168908903939.

[10.11] 'Change detection techniques', D Lu, P Mausel, E Brondiacutezio and E Moran, *International Journal of Remote Sensing*, Vol. 25(12) (2004), pp.2365–2407.

[10.12] 'Remote sensing of forest health', J Tuominen, T Lipping, V Kuosmanen and R Haapanen in *Geoscience and Remote Sensing* (Intech, 2009), pp.29–52, ISBN 9789533070032.

[10.13] www.trafficengland.com/

[10.14] www.farsondigitalwatercams.com/

[10.15] www.maineharbors.com/camindex.htm

[10.16] www.chamonix.net/english/webcams/chamonix-town

[10.17] www.chamonix.com/webcam-chamonix-brevent-altitude-2000,139,fr.html

[10.18] www.dartmoorcam.co.uk/cam/Links/Links.htm

[10.19] www.dartcom.co.uk/weather

[10.20] www.livefromiceland.is/webcams/reykjavikurtjorn/

[10.21] support.esri.com/en/technical-article/000011244

[10.22] www.qgis.org/en/site/

[10.23] www.bilko.org/

APPENDIX 1: NUMERICAL SOLUTIONS

Chapter 1

[1.4] Reflected power = $\rho(\lambda) \times \Phi(\lambda) = 0.4 \times 100 = 40W$

Transmittance $\tau = \Phi - \rho - \alpha = 1 - 0.4 - 0.2 = 0.4$

Transmitted power = $\tau \times \Phi(\lambda) = 0.4 \times 100 = 40W$

[1.6] $\lambda_{max} = 2897/T$ in microns $T = 2897/\lambda_{max} = 2897/9.73 = 297.7K = 24.6°C$

[1.7] $W_T = \sigma T^4$, $W_{1.1T} = \sigma(1.1T)^4$, $W_{1.1T}/W_T = 1.464$ approx. 46.4% radiated intensity increase

[1.8] UV dose = $280 \times 1 \times 60 \times 60 = 1.008 MWm^{-2}s^{-1}$

[1.9] $0.59 Wm^{-2}$

Chapter 2

[2.1]

Transmittance $\tau = \Phi - \rho - \alpha = 1 - 0.1 - 0.4 = 0.5$

Reflected power = $\rho(\lambda) \times \Phi(\lambda) = 0.1 \times 100 = 10W$

Absorbed power = $\alpha(\lambda) \times \Phi(\lambda) = 0.4 \times 100 = 40W$

[2.3] For soil the transmittance = 0 so $\Phi = \rho + \alpha$ so $\rho = \Phi - \alpha = 1 - 0.3 = 0.7$

[2.7] Roughness = $\lambda/(8\cos\theta) = 2.5cm$. Mean surface roughness = 2cm so the surface is smooth

[2.9] Loam reflectance = 0.2, peat reflectance = 0.08, ratio = 2.5, reflectance = 0.45 − 0.3 = 0.15, or in dB −8.3dB

[2.10] $L(R) = 1/452 \, Wm^{-2} \, T = 54\%$

Chapter 3

[3.1] Using $M_{0.5microns,sun} = \dfrac{3.74151\times 10^8}{5^5} \dfrac{1}{e^{\frac{14387.9}{5\times 6000}} - 1} \, Wm^{-2}\mu m^{-1}$

$M_{5microns,sun} = 0.195 M \, Wm^{-2}\mu m^{-1}$

Using $T_{rad} = \sqrt[4]{\varepsilon T_{kin}^4 + \rho T_{sky}^4} = \sqrt[4]{0.2(273+30)^4 + 0.3(273-40)^4} = 225K$

[3.2] Material 1 $ATI = \dfrac{1-0.8}{22-8} = \dfrac{0.2}{14} = 0.0143 K^{-1}$

Material 2 $ATI = \dfrac{1-0.8}{28-6} = \dfrac{0.2}{22} = 0.009 K^{-1}$

[3.3] $I_{man} = \sigma \varepsilon T^4 = 0.8\sigma(9273 + 23.5)^4 \quad I_{sea} = 0.9\sigma(273+9)^4$

Relative difference $(I_{man} - I_{sea})/I_{man} = 6.4\%$. The man is **not** discriminated!

Crossover occurs where: $\varepsilon_{man}\sigma T_{man}^4 = \varepsilon_{sea}\sigma T_{sea}^4$ Hence $T_{man} = 297K$

[3.4] $T_{0.8} = 283.7K, T_{0.5} = 252.3K$

Peak radiated wavelength $= \dfrac{2900}{6900} = 0.42 \, microns$

[3.5] $T_2 = (SST(True)-AT_1-C)/B = (320-0.88 \times 310-18)/0.15 = 248K$

[3.6] Bands: (a) 5, 7; (b) 3, 4; (c) 1, 2

[3.7] (a) total emissive power $= 2.145 \, Wm^{-2} nm^{-1}$

(b) Peak radiated wavelength $= \dfrac{2900}{T} \quad T = \dfrac{2900}{wavelength} = \dfrac{2900}{300} = 9.7 \, microns$

[3.9] (i) (a) 11.24 (b) 2.024 and (c) 9.03 microns

(ii) Arctic snowfield 236Wm^{-2}, molten lava 196kWm^{-2} and hot desert 258.9Wm^{-2}

[3.10] 301.5K

$P = (k.\rho.c)^{1/2} = (1.13 \times 2000 \times 2000)^{1/2} = 2126$ Jm^{-2}K^{-1}s$^{-1/2}$

Chapter 4

[4.3] Rr = cτ/(2sinθ) Ra = Hλ/(Lcosθ) (a) Ra = 106m, Rr = 6.4m (b) Ra = 382.4m, Rr = 8.85m (c) Ra = 3.163km, Rr = 0.0037m (d) Ra = 3.3km, Rr = 0.0186m

[4.4] Horizontal beam width = 0.75°

[4.5] 375m

[4.6] $α_H = 60λ/[(n-1)d] = 60λ/[(30-1)0.5] = 4.14°$

[4.7] SAR = VT = 7900 × 2.5 = 19.75km

[4.8] $\Delta f = \dfrac{2 \times 3.1 \times 10^9 \times 288}{3 \times 10^8} = 5952.0 \text{Hz}$

[4.9] $\phi = 360 \times 0.5 \times \sin 60 = 155.8°$

Chapter 5

[5.2] $p_{water} = p_{partial}$ = totalpressure water vapour volume fraction

$$H = \dfrac{p_{water}}{p_{sat}(T)} = \dfrac{101 \text{water vapour volume fraction kPa}}{2.57 \text{kPa}} = 0.2$$

Water vapour volume fraction = 0.005

[5.4] $p = p_0 e^{-\frac{gM}{RT}h} = 39200 \text{Pa}$

[5.5] $p = p_0 e^{-\frac{gM}{RT}h} = 106631.9 \text{Pa}$

[5.6] $I_{scattering}$ proportional to inverse fourth wavelength, Scattering ratio 400/600nm = 5.063

Chapter 6

[6.4] See [3.5] $T_2 = 345K$

[6.5] 615km, energy = 3.7×10^{-19}J

[6.6] Emissivities 0.9 and 0.6 respectively yield temperatures of 295K and 266.7K respectively

Chapter 7

[7.4] Montmorillonite and illite reflectance ratio approx. 2 at 1 microns, and 1.8 at 1.4 microns

[7.6] Ratios Fe/Mn a) 400; b) 0.025; c) 500. If Earth's crust has an Fe/Mn ratio = 0.53, 0.0053%, the likely crustal deposit is b)

Chapter 8

[8.1]

$$n_p = \sqrt{1-\left[\frac{f_p}{f}\right]^2} = \sqrt{1-\left[\frac{10}{15}\right]^2} = 0.745$$

[8.2] $\Delta t = \dfrac{2h}{c} = \dfrac{2 \times 85000}{c} = 0.57$ms

[8.3] $f_p = \sqrt{\dfrac{Nq^2}{4\pi^2 \varepsilon_0 m}} = \sqrt{\dfrac{10^{11} \times (1.6 \times 10^{-19})^2}{4\pi^2 \times 8.84 \times 10^{-12} \times 9.9 \times 10^{-31}}} = 2.72$MHz

[8.4] $N_{max} = 1.24 \times 10^{10} \times 12^2 = 1.79 \times 10^{12}$ electrons m^{-3}

[8.6] $n_p = \sqrt{1-\left[\dfrac{f_p}{f}\right]^2}$ $f_p = \sqrt{81 N_{max}}$

So $n_p = \sqrt{1 - \left[\frac{81 N_{max}}{1.8 kHz}\right]^2}$

$n^2 - 1 = -\frac{81 \times 3.24 \times 10^4}{3.24 \times 10^6}$ $n = 0.44$

[8.10] Use: $p = p_0 e^{-\frac{gM_h}{RT}}$ $p = 6 \times e^{-\frac{3.7 \times 0.08944}{8.314 \times 220} \times 100} = 0.11 mbar!$

Chapter 9

[9.1] $gt = \frac{g_0 R^2}{(R+H)^2} = \frac{9.81 \times 6400^2}{(6400+1200)^2} = 0.709 g_0$

[9.2] $mgt = \frac{mg_0 R^2}{(R+H)^2} = \frac{mv^2}{(R+H)}$ so $\frac{g_0 R^2}{R+H} = v^2$ $v = \sqrt{\frac{g_0 R^2}{R+H}} = 7576 ms^{-1}$

[9.4] $V_{circular} = (gR)^{1/2}$ R~6380km so $V_{circular}$ ~7.9ms^{-1}

Due to gravitation, the circular velocity required to keep spacecraft in orbit decreases as distance from Earth increases.

[9.5] $W = \int_{Ra}^{Rb} \frac{GMm}{R^2} dr$ $W_b - W_a = \int_{Ra}^{Rb} \frac{GMm}{R^2} dr = -GMm \left[\frac{1}{Rb} - \frac{1}{Ra}\right]$

$W = 0$ at $R = \infty$ $W_b = -\frac{GMm}{Rb}$ from potential energy considerations for escape $\frac{mV^2}{2} = \frac{GMm}{Rb}$

[9.6] $\frac{mV^2}{2} = \frac{GMm}{Rb}$ $V^2_{escape} = 2GM/R$ Thus: $V_{escape} = (2GM/R)^{1/2}$ $GMm/R^2 = mg$, as GM/R
$= g$ $V_{escape} = (2GR)^{1/2}$ $g \approx 9.81 ms^{-2}$ $R \approx 6,400 km$ so $V_{escape} \approx 11.23 kms^{-1}$ in **any** direction!

[9.7] Initial KE and PE = Final KE and PE

$\frac{mVinitial^2}{2} - \frac{GMm}{11Re} = \frac{mVfinal^2}{2} - \frac{GMm}{Re}$

$$V_{final}^2 = \frac{2GM}{R_e}\left[1-\frac{1}{11}\right] + V_{initial}^2$$

So $V_{final} = 16.8 \text{ kms}^{-1}$

[9.8] $7500 = 2300\ln(100/(100-M_f))$ so $M_f = 96.17\%$

[9.9] Taking a circular orbit, $F = mv^2/R = MmG/R^2$ $v^2 = GM/R$ but $v = R\omega = R(2\pi f) = 2\pi R/T$ Hence $V^2 = 4\pi^2 R^2/T^2$ and $4\pi^2 R^2/T^2 = GM/R$ so $R^3 \propto T^2$, Kepler's third law

Increase of $\sqrt{8}$ in period for doubling range

[9.10] $(R+H) = \frac{GM}{V^2}$ so $H = 200 \text{km}$ $F = \frac{mV^2}{(R+H)} = 27.4 \text{N}$

Using $T^2 = 4\pi^2 R^3/GM$ $T = 5335\text{s}$ or 88.9 minutes

Chapter 10

[10.6] $k = 2$ so $L = 2k = 4$, $r = 2$, so the transformation $s = (L - 1) - r$ is applied, $s = (4 - 1) - 2 = 1$

GLOSSARY

Ablation Loss of ice and snow from a glacier through melting and run-off, sublimation, evaporation, calving etc.

Absolute Zero The lowest temperature possible on the Kelvin scale.

Absorbance (or absorptance) Ratio of the radiant energy absorbed by a target to the energy incident on it.

Absorption band A wavelength range where electromagnetic radiation is absorbed by the atmosphere.

Accumulation Addition of ice and snow into a glacier, eg through precipitation.

Active remote sensing Remote sensing methods that use their own source of electromagnetic radiation to illuminate a target surface.

Albedo The reflectivity of a surface. Ice has a high albedo, meaning solar radiation is mostly reflected. Dark ocean has a low albedo, so solar radiation is mostly absorbed.

Apogee The highest point in the orbit of a satellite.

Apparent Thermal Inertia An approximation of thermal inertia calculated as 1 – albedo divided by the difference between day and night radiant temperatures.

Atmospheric correction Image processing method that compensates for a particular problem, eg the effects of selectively scattered light due to the atmosphere.

Atmospheric window Wavelength range where the atmosphere readily transmits electromagnetic radiation.

Azimuth Geographical orientation of a line given as an angle measured in degrees clockwise from north.

Biosphere The regions of the Earth surface and atmosphere occupied by living organisms.

Black body A theoretical perfect absorber and emitter of energy.

C-band Radar frequencies from 4 to 8GHz.

Calving A process by which pieces of ice break away from the terminus of a glacier that ends in a body of water, or from the edge of a floating ice shelf that ends in the ocean.

Calibration A process where an instrument's measurements are compared with a standard or reference measurement, usually 'ground-truthed' on the Earth.

Carbon cycle Describes the movement of carbon through the environment. Carbon exists in the ground, sea, in organisms and the atmosphere.

Change detection imaging Differences between images, usually taken on different dates, which are compared to look for changes.

Cryosphere The frozen water part of the Earth system.

Dielectric constant The electric property that affects the interaction of electromagnetic energy with matter.

Electromagnetic spectrum Continuous sequence of electromagnetic energy.

Emissivity The ratio of the energy radiated from a material surface to that radiated for a perfect black body emitter.

ESA European Space Agency.

Exitance The outgoing flux of radiant energy per unit area.

Exosphere The outermost region of the Earth's atmosphere.

Feedbacks Processes in the Earth system that change as a result of a change, such as the albedo of a surface.

GEO Geostationary Earth Orbit.

GIS Geographical Information System.

Greenhouse gases Any gas that absorbs and emits thermal radiation, contributing to the greenhouse (blanket) effect.

Hydrosphere All the waters on the Earth's surface, such as seas and lakes.

Hyperspectral System that collects and processes information across hundreds of individual wavelength bands.

I-band Radar frequencies from 8 to 10GHz.

IFOV Instantaneous Field Of View.

Ice sheet Ice masses covering over 50,000km^2.

Insolation Incoming Solar Radiation.

Ionosphere The region of the atmosphere between 80 and 1,000km.

Ionosonde A special radar used to examine the ionosphere.

Irradiance The incoming flux of radiant energy per unit area.

Isotope Two or more forms of an element with different numbers of neutrons in the nucleus of an atom. This makes some atoms lighter or heavier than others, despite being the same element.

ISAR Inverse Synthetic Aperture Radar.

Kinetic temperature The actual temperature of an object, measured with a contact thermometer.

L-band Radar frequencies from 1 to 2GHz.

Layover In radar images, the geometric displacement of the top of an object towards the near range relative to its base.

LEO Low Earth Orbit.

LIDAR LIght Detection And Ranging.

Mass balance A measure of the health of a glacier (inputs minus the outputs).

Mesosphere The region of the atmosphere between 50 and 80km.

MEO Middle Earth Orbit.

Model A representation of reality, usually made from a combination of real-world observations and mathematical calculations to predict past or future change.

Mosaic A composite image created by combining a large number of overlapping images.

Multispectral System that collects and processes information across several individual wavelength bands.

Nadir Point on the ground directly below the remote sensing system towards the centre of the Earth.

NASA National Aeronautics and Space Administration.

NDVI Normalised Difference Vegetative Index.

Oblique photographs Photographs acquired with a camera at a large angle between the horizontal and vertical directions.

Panchromatic film Black-and-white images acquired across all visible wavelengths.

Passive remote sensing Remote sensing methods that use naturally reflected radiation to illuminate a target surface.

Perigee The lowest point in the orbit of a satellite.

Permafrost Frozen soil that survives the summer melt Permafrost can be continuous in extent (covering > 90%

of an area), discontinuous (50–90%) or sporadic (10–50%).

RADAR Radio Aid for Detection And Ranging.

Radar shadow Topographical features that block incident energy create dark shadows on the radar image.

Radiant temperature Concentrated radiant flux from a material.

Rayleigh criterion The relationship between surface roughness, wavelength and incident angle that determines whether a surface appears rough or smooth to the incident radiation.

Reflectance Ratio of the radiant energy reflected by a target to the energy incident on it.

Remote sensing The science of acquiring, processing and interpreting images that record the interaction between electromagnetic energy and the Earth-atmosphere system.

S-band Radar frequencies from 2 to 4GHz.

Scanning system An imaging system in which the IFOV of one or more detectors is swept across the illuminated surface.

Scattering Refers to the reflections of electromagnetic waves from multiple particles or surfaces.

Scattering Multiple reflection of electromagnetic waves by particles or surfaces.

SST Sea Surface Temperature.

SWIR Short Wave Infra red.

SLAR Side Looking Airborne Radar.

Slant range A direct line running between the receive antenna and the illuminated surface.

Speckle Grainy noise that degrades the quality of the radar image, caused by coherent processing of backscattered signals from multiple distributed targets.

Spectral resolution The range of wavelengths recorded by a detector.

Spectral sensitivity The response of the detector system to radiation across different spectral regions.

Spectrometer A detector that measures radiation intensity as a function of wavelength.

Spectrum The continuous sequence of electromagnetic energy arranged according to either frequency or wavelength.

Supervised classification Information extraction technique in which the human operator provides training-site information, which computer software then uses to assign pixels into classification categories.

Stefan-Boltzmann constant The Stefan-Boltzmann constant has a value of approximately 5.67×10^{-8} Wm^{-2} K^{-4}.

Stefan-Boltzmann law The radiant flux of a black body is equal to the temperature to the fourth power times the Stefan-Boltzmann constant.

Stratosphere The second major atmospheric layer above the troposphere, extending from 10 to 50km altitude above the Earth.

SAR Synthetic Aperture Radar.

Thermal conductivity Measure of the rate at which heat passes through a material.

Thermosphere The region of the atmosphere where the atmosphere ceases to have the properties of a continuous medium.

Transmittance Ratio of the radiant energy transmitted by a target to the energy incident on it.

Unsupervised classification Information extraction technique in which computer software alone assign pixels into classification categories.

Wavelength The distance between successive wave crests or troughs.

Wien's displacement law Describes the shift of the radiant power peak to shorter wavelength as temperature increases.

X-band Radar frequencies from 8 to 12GHz.

X-ray A range of electromagnetic energy having wavelengths 0.01–10 nm, corresponding to frequencies 3×10^{16}–3×10^{19} Hz, and energies in the range 100eV–100keV.

INDEX

absorbance a 25, 125
absorbance, radiation 23–4, 42–3, 48, 59–60, 67, 116, 117–18, 125, 126–7, 128, 129, 188
ACE (Advanced Composition Explorer) satellites 176, 179, 186
Active Microwave Instrument (AMI) 109–10
ADEOS (ADvanced Earth Observing Satellite) 176, 210
aerial/airborne platforms *see* aircraft; balloons; UAVs
aerial photography 10, 155, 158, 160, 167, 194, 208
aerosols 4, 17, 127, 142, 175, 178, 186, 187, 188, 189–90
AESA (Active Electronically Scanned Arrays) 97, 106–7
agriculture/farming 8, 10, 38, 52, 77, 162, 163 *see also* vegetation
aircraft 1, 7, 10, 81, 82, 97, 140, 155, 156, 165, 168, 194–5
AIRS (Atmospheric InfraRed Sounder) 176, 206
albedo 23–4, 58–9
algae 18, 20, 41, 53, 133–4, 137
altimeters, radar 107, 109–10, 111, 113, 145, 149, 150, 206, 207
AMSR (Advanced Microwave Scanning Radiometer) 82, 206
AMSU (Advanced Microwave Sounding Unit) 201, 206
antennas, radar 95–6
apogee and perigee altitudes 200–1
Apparent Thermal Inertia (ATI) 68–9
Appleton, Edward 181–2
Aqua satellites 18, 131, 137, 150, 176, 196, 206
Aquarius radiometers 82, 150, 211
Aral Sea, Kazakhstan 134–5
archaeology 19, 76–7, 164–6
ASAR (Advanced Synthetic Aperture Radar) 206
ASTER (Advanced Spaceborne Thermal Emission and Reflection) 19, 74, 76, 77, 146, 156, 166, 167, 170

atmospheric interactions
 absorbance and transmittance 125
 atmospheric absorption spectrum 117–18
 atmospheric and ionospheric turbulence 123
 atmospheric composition 120–1
 atmospheric transmission 118–19
 ocean attenuation 125–7
 ozone layer 127
 radiation from Earth 119–20
 radiation from the sun 116–17
 radiation propagation 123–4
 remote sensing inverse problem 128–9
 and the solar radiation spectrum 116–17
atmospheric remote sensing applications 1, 175–7
 atmospheric layer sensing - exosphere 179–81
 measurement geometries 178–9
 satellite-based sensing - stratosphere 185–6
 satellite validation principles 190–1
 space-based measurements - troposphere 188–9
 space-based models - mesosphere 184–5
 space-based models - thermosphere 184
 terrestrial measurement - ionosphere 181–4
 terrestrial measurements - troposphere 186–8
ATSR (Along Track Scanning Radiometer) 77, 146, 170, 188–9, 206
attenuation, atmospheric 22–4, 42–6, 116, 119, 124, 125–6
Aura satellites 176, 185, 186, 189, 206
AVHRR (Advanced Very High Resolution Radiometer) 54, 77, 78–9, 82, 91, 119, 144, 146, 148, 160, 168, 170
AVIRIS hyperspectral imager 22, 160, 162, 167
azimuth, solar and sensor 52

backscattered signals 31, 101, 111, 113, 137, 148–9, 170
balloons 10, 194
beam transmissometers 45

Beer-Lambert law 43–6
bispectral sensing 22
black bodies/black body radiation 27, 28, 64–6
blue band 17

CALIPSO 176, 189, 210
Cassegrain antennas 96
Celsius scale 26
CERES (Clouds and the Earth's Radiant Energy System) 189, 206
chlorophyll 18, 19, 41–2, 48–50, 73–4
Cirrus cloud band 19
classification, remote sensing system 12–13
Closed Circuit TV (CCTV) 15, 231
clouds/cloud cover 4, 15, 17, 19, 20, 47, 56, 75, 90, 91, 92, 106, 111, 113, 119, 123, 125, 144, 145–6, 155, 170, 176, 186, 187, 188, 189, 202, 206
CloudSat 176, 189, 211
Cluster satellites 117, 179, 180
CNES 205, 207
coastal aerosol band 17
cooling photon detectors 81–2
Copernicus 145, 175–6
coronal mass ejections (CMEs) 121, 180
CryoSat 146, 150, 151, 209
cryosphere 4, 8, 20, 131–2
  glaciers and icebergs 74, 75, 113, 134, 144, 146, 166–7
  ice shelves 134
  and microwave radiation 111, 113–14
  and NIR and visible radiation 58–9
  passive radar use 146
  sea and ice radar interferometry 149–50
  sea snow and ice mapping 146
  and thermal infrared radiation 74–5
  and ultraviolet 59–60

desert sands 156
desertification 170
detectors, IR 83–5
DIAL (Differential Absorption Lidar) 185, 187, 188
dielectric properties, mineral 111, 113, 160
disease, geospatial mapping of 170
dish antennas 95–6
Dobson units 48
Doppler radar 95, 103, 105, 106–7, 109, 178

DORIS (Doppler Orbitography and Radio Integrated by Satellite) 206, 207, 209

Earth observation overview 1–4
  historical review 9–11
  requirements in the 21st century 6–9
earthquake radar interferometry 170–1
echo ranging principle 92
electromagnetic radiation overview 12–21
emissivity, object 20, 65–9, 70–1, 74–5, 91, 111, 113, 146, 148
Envisat satellites 105, 176–7, 185, 189, 194, 205–6, 208
EOS (Earth Observing System) 77, 150, 206
ESA 86, 105, 109–10, 146, 180, 185, 194, 207–8
European Remote Sensing (ERS) satellites 86, 107, 109–11, 127, 146, 149–50, 166, 171, 176, 177, 188–9
exitance (emittance) 27–8
exosphere 177, 179–81

false-colour infrared film 11
FIR bands and sensors 16, 82–3, 119, 148
flood monitoring systems 138–9
fluorescence, rock and mineral 58
Flux Incident 36
forestry 8, 10, 38, 52, 78, 112, 162, 163–4, 170
  see also vegetation

GALE (Giant Aperture Lidar Experiment) 188–9
gamma rays 14
geological mapping 154
geology *see* rocks and minerals
geophysical air surveys 156
Geostationary Operational Environmental Satellite (GOES) 20, 202
geostationary satellites 201–2, 203
Giovanni data resource, NASA 190
glaciers 74, 75, 134, 146, 166–7
GNSS (Global Navigation Satellite System) 9, 181
GOME (Global Ozone Monitoring Experiment) 127, 176, 177, 185, 186
GOMOS (Global Ozone Monitoring by Occultation of Stars) 176–7, 206
GPS (Global Positioning System) 9, 123, 140, 178, 181

greenhouse gases 175, 176–7, 187
Gregorian antennas 96
grey bodies 65, 70
ground heating rate 72

HCMM (Heat Capacity Mapping Mission) 76, 79, 82, 85, 168
heat capacity c 68
hemispherical reflectance 29–30
humanitarian disasters 8–9, 171
hydrosphere 131–2
   lake eutrophication 133–4
   and microwave radiation 111, 113–14
   and NIR and visible radiation 41–2
   ocean attenuation 125–6
   ocean colour 136–7
   ocean wind measurement 137–8
   oil spillages 149
   rivers and remote sensing 138–9
   sea and ice radar interferometry 149–50
   sea snow and ice mapping 145
   sea surface height (SSH) 145
   sea surface temperature (SST) 72–3, 144
   space-based active radar 143–6
   space-based visible applications 142
   terrestrial maritime thermography 140–1
   terrestrial maritime visible imagery 140
   and thermal infrared radiation 72–3
   and ultraviolet 47–8
   underwater light attenuation 42–6
   water pollution detection 132–3
   water properties 38–40
   water security issues 134–5
   water vapour detections channels 20
   wetland mapping 139–40
Hyperion hyperspectral imagers 22
hyperspectral imagers 22

ice *see* cryosphere
icebergs 113, 134, 144
IFOV (Instantaneous Field of View) 79, 80–1
IKONOS 21, 55, 165, 166, 167, 211
infrared radiation 9, 15
infrared radiometers 79–81
insolation 23, 24, 72, 116–17
interferometry 109, 111, 118, 149–50, 159, 167, 169, 170–1, 176, 206
International Space Station 184, 196, 207
inverse problem, remote sensing 128–9

Inverse Synthetic Aperture Radar (ISAR) 109
ionosphere 121, 181–4
IRLS (InfraRed Line Scanners) 82
irradiance 27

Jason satellites 101, 145, 149, 150, 180, 207, 208

Kelvins 26
Kepler's laws 199–201
kinetic temperature 70–1
Kirchhoff's law 28, 67, 74

lake eutrophication 133–4
Lambertian surface variation 37
land glaciers 4, 166–7
   see also cryosphere
land resource remote sensing applications 153
   archaeology 164–6
   disaster monitoring 170–1
   forest applications 163–4
   humanitarian applications 171
   land glaciers 166–7
   land surveillance 169
   land use/cover mapping classification 160–1
   monitoring urban growth 167–8
   rocks and minerals 154–8
   soil 158–60
   terrestrial building heat surveys 168
   urban and regional planning applications 167
   vegetation cover 162–4
Landsat satellites 11, 17, 20, 22, 37, 52, 54, 55, 76, 77, 78, 82, 113, 123, 133, 142, 144, 148, 153, 155, 156, 160, 165, 166, 167, 168, 170, 204–5
Leaf Area Index (LAI) 18, 50
lidar (Light Detection and Ranging) 21, 83, 147–8, 167, 185, 187–8
limb sounding 176, 178, 186
LITE (Lidar In-space Technology Experiment) 188

magnetosphere 179, 180
man-made disasters 147–9, 170–1
maritime applications 75, 91–2, 101, 103, 105–7, 109, 113, 140

oil spillages 147–9
passive radar 145–6
sea and ice radar interferometry 149–50
sea surface height (SSH) measurement 145
sea surface temperature (SST) measurement 144
space-based radar 143–6
space-based visible 142
terrestrial maritime thermography 140–1
terrestrial maritime visible imagery 140
MERIS (MEdium Resolution Imaging Spectrometer) 18, 42, 206
mesosphere 121–2, 177, 184
meteorological applications 8, 9, 11, 97, 118, 144
see also weather predictions and systems
Meteosat satellites 19, 118, 148, 202, 203–4, 208
methane emissions 175
microwave radiation and remote sensing 11, 16–17
  active sensors 91–2, 109–10, 165–6
  brightness temperature 90
  interaction with rocks and minerals 111, 112–13
  interaction with snow and ice 111, 113–14
  interaction with soil 111, 112
  interaction with vegetation 112, 162
  interaction with water 111
  ocean wind measurements 137–8
  passive sensors 90–1, 113, 145–6
  problems with visible imagery 90
  radiation from Earth 119–20
  sensor system types 109–11
  see also radar
Mie scatter 125
military applications 9, 10, 11, 12, 15, 97, 106–7, 109, 140, 141, 165, 169–70
  see also surveillance maritime applications
minerals and rocks *see* rocks and minerals
MIPAS (Michelson Interferometer for Passive Atmospheric Sounding) 176, 177, 185, 186, 206
MIR bands and sensors 15, 85–6, 119
MODIS (MOderate Resolution Imaging Spectroradiometer) 18, 37, 42, 77, 82, 137, 138, 144, 146, 148, 160, 170, 206

MODTRAN (MODerate resolution atmospheric TRANsmission) 179
MOPITT (Measurements Of Pollution In The Troposphere) 167, 176, 190

nadir sounding 80, 111, 127, 178, 185, 186, 207
nano-satellites 208–9
narrow-waveband sensing 22
NASA 9, 22, 54, 138, 159, 165, 167, 176, 179, 180–1, 185, 190–1
natural disasters 7, 9, 138–9, 170–1, 195
Near InfraRed (NIR) and remote sensing 15, 18, 19, 21
  archaeology 165
  desert sands 156
  and the hydrosphere 41–2
  oil spillages 148
  and rocks and minerals 58, 155
  and snow and ice 58–60, 166
  and soil 55–6, 160
  urban areas 168
  and vegetation 48, 50–1, 53, 54, 162, 164
Nimbus satellites 22, 80, 118, 119, 146, 176, 185, 194
NOAA 20, 54, 78–9, 91, 118, 179, 201
non-Lambertian surface variation 37
non-selective scatter 125
Normalised Difference Vegetation Index (NDVI) 18, 19, 53–5, 112, 162, 164, 170, 171

occultation 178
oceans and seas 8, 38–9, 70
  attenuation 125–6
  measuring salinity 91, 146
  ocean colour 136–7
  ocean wind 137–8, 146
  radar imaging 101, 113–14
  sea ice 4, 64, 75, 113, 145, 146
  sea surface temperature (SST) 72–3, 144, 146
  see also hydrosphere; maritime applications
Odin satellites 176, 186
oil exploration programmes 156, 158
oil spillages 147–9, 207
Operational Linescan System, DMSP satellite 168

optical refraction under water 46
ozone 15, 47–8, 127, 175, 176, 185, 186, 187, 188, 189–90, 194, 206

pan-sharpened systems 21
panchromatic band 20
panchromatic systems 21, 158, 160, 165
parabolic reflectors 95–6
PARASOL 176
passive microwave sensors 90–1, 113, 145–6
perigee and apogee altitudes 200–1
phased radar arrays 96–100
phenological cycles 53
photodetectors, IR 85–6
photon detectors 81–2
phytoplankton 136–7
Planck's black body equation 27, 64
Pleiades satellites 166, 207, 210
Polar orbiting satellites 201, 202–3
polarisation, radar image 109, 111, 112, 113
pollution, water 132–3, 134–5, 149
Poseidon radar altimeters 107, 145, 149, 207
propagation, radiation 22–4
pulse delay ranging 92
pyroelectric IR detectors 84–5

QWIP (Quantum Well Infrared Photodetectors) 85–6

radar 10, 11, 17, 75, 91–2
　altimeters 109–10, 111, 113
　angular resolution 93–4
　antennas 95–6
　archaeology 165–6
　ASAR 206
　Doppler radar 95 (see also SAR)
　imaging 100–6, 166
　ISAR 109
　oil spillages 148–9
　phased arrays 96–100
　polarisation and 3D images 109, 111
　range resolution 94
　SAR 103–8, 141, 143, 149, 155, 156, 160, 165, 167, 169–79, 207
　sea and ice radar interferometry 149–50
　SLAR 101–3

soil salinity 160
urban mapping 168
see also microwave radiation and remote sensing
RADARSAT satellites 112, 113, 204
radiant temperature 67, 68–9, 70
radiation emissions 26–9
radiation wavelengths 14–17
radio waves 9, 17
radiometers, microwave 76–7, 79–81, 83, 85, 90–1, 146, 206, 207
radiometric corrections 216
radiometric resolution issues 22
Rayleigh's criterion 37, 43, 124–5, 188
red band and red edge band 18, 164
reflectance p 25, 67–8
reflectance, radiation 15, 42, 48, 50–1, 52, 53, 55–6, 60, 119, 162
reflected near infrared radiation 15
remote sensing systems 1–4
　application sensing bands 17–20
　classification of 12–13
　components and process 30–2
　data acquisition, analysis and utilisation 3–4
　inverse problem 128–9
　non-imaging 12, 79–82
　passive 12, 23, 116
　radiant temperature 67
　radiation sources 20–1
　resolution issues 22
　types of imagery 21–2
　see also atmospheric interactions; atmospheric remote sensing applications; land resource remote sensing applications; maritime applications; Near InfraRed (NIR) and remote sensing; radar; satellites; thermal infrared radiation (TIR) and remote sensing; ultraviolet radiation and remote sensing; visible radiation waves and remote sensing
rescue forces 7, 16, 140, 169–70
resolution issues 22, 37, 167
rivers and remote sensing 138–9
rocketry 195–6
rocks and minerals 8, 19, 20, 77, 113, 154
　and microwave radiation 111, 112–13
　remote sensing applications 154–8

Index

thermal infrared radiation 76
and visible, NIR and UV radiation 58
roughness, surface 37, 41–2, 52, 56, 91, 113, 137, 146

satellite image processing 215
  atmospheric corrections 216–17
  change detection 228–9
  DN values 216
  geometric corrections 215–16
  image classification and analysis 229–30
  image enhancement 217–23
  image interpretation 227–8
  image transformations 223–7
  matrix filter approach 219–23
  'noise' on images 217
  pre-processing 215–17
  radiometric corrections 215
  scattering 216–17
  spatial filtering 219
satellites 1, 7, 9, 11
  apogee and perigee altitudes 200–1
  circular orbit and velocity 198
  early history of 194–5
  Earth observation satellites 204–12
  effect of ionosphere 121, 124
  elliptical orbits and escape trajectories 198–9
  geostationary satellites 201–2, 203
  high resolution satellite missions 208
  Kepler's laws 199–201
  physics basics 197–201
  Polar orbiting satellites 201, 202–3
  remote sensing inverse problem 128–9
  rocketry 195–6
  small satellite missions or nanosats 208–9
  temperature of 66
  thermal IR systems 76–9
  thermal non-imaging systems 79–82
  thermal scanners 82–3
  types of satellites 201–12
  validation principles 190–1
  weather sensing satellites 203–4
  web-based image sources 231–3
  what is a satellite? 196–7
  see also remote sensing systems; individual satellites by name
scanners, thermal 82–3, 168
scanning sensors 85

scattering 23–4, 42–3, 59, 67, 75, 102, 109, 111, 116, 117, 118, 119, 123–5, 126–7, 128, 129, 188
SCIAMACHY (SCanning Imaging Absorption spectroMeter for Atmospheric CHartography) 177, 185, 189–90, 206, 208
SCISAT ACE 186
sea ice 4, 64, 75, 113, 145, 146
Sea Surface Temperature (SST) 72, 144, 146
Sea-viewing Wide Field-of-view Sensor (SeaWiFS) 18, 42, 43, 51, 136–7
SeaSat 105, 146, 165, 166
SeaSpray radar 106–7
security issues, water 134–6
seismic surveys 156, 158
sensor elevation angle 51–2, 74
Sentinel satellites 81, 145, 150, 161, 186, 189, 207–8
Short Wave InfraRed 1 SWIR1 and 2 19–20
Shuttle Imaging Radar (SIR) 107
Side-Looking Airborne Radar (SLAR) 10, 101–3, 107–8
6SV method, inverse problem and the 129
snow 4, 20, 55, 91, 111, 113–14, 123, 145, 162
  see also cryosphere
social media/smart phones 231
soil 18, 19, 50, 51, 52
  desert sands 156
  mapping and evaluation 158–9
  and microwave radiation 111
  moisture, structure and texture 56, 146
  and NIR and visible radiation 55–6
  organic matter and iron oxide 56
  salinity 159–60
  Soil Thermal Inertia (STI) 75
  and thermal infrared radiation 75
  and ultraviolet 56
solar absorbance 66–7
SOlar and Heliospheric Observatory (SOHO) 14, 117, 179, 180, 196
solar elevation 51
solar energy 31, 65
solar heating cycle 69, 70, 72
solar radiation 23–4, 25, 28, 29, 54, 66, 70, 116–17
solar reflectance 66
space-borne thermal imagers 77–9
space weather 179–80

spectral radiant flux 24, 25, 36–7, 125
spectral resolution issues 22
spectrometers, thermal 79, 81, 83
SPOT satellites 52, 123, 166, 205, 207
spotlight, SAR 106
Sputnik 196
staring arrays, thermal sensor 83
Stefan-Boltzmann law 27, 64
stratosphere 121, 177, 185–6, 187, 188
Surface Composition Mapping Radiometer (SCMR) 80–1
surface variation 37, 41–2
surveillance maritime applications *see* maritime applications
suspended materials 41–2
swarm satellites 181
swath mapping, SAR 106
Synthetic Aperture Ladar (SAL) 169–70
Synthetic Aperture Radar (SAR) 10, 103–9, 113–14, 138, 141, 143, 149, 155, 156, 160, 165, 167, 169–70, 207

temporal resolution issues 22
Terra satellites 77, 131, 150, 176, 190, 206
TerraSAR satellites 114, 207
terrestrial building heat surveys 168
Thematic Mappers (TM) 77–8
thermal
  radiometers 76–7, 79–81
thermal conductivity k 68
Thermal Imaging Cameras (TIC) 16, 75, 82
thermal inertia 68–9, 76–7, 165
thermal infrared radiation (TIR) and remote sensing 15–16
  archaeology 165
  and the cryosphere 74–5, 166
  and the hydrosphere 72–3
  oil spillages 148
  and rocks and minerals 76
  and soil 75, 160
  urban areas 168
  and vegetation and chlorophyll 73–4, 162
Thermal InfraRed Sensor TIRS1 20
thermal radiation and remote sensing
  advanced cooled FPAs 85–6
  and the Earth's surface 64–5, 72–6
  emerging un-cooled FPAs (UFPA) 86
  far infrared thermal imaging sensors 82–3
  IR detectors 83–5

kinetic temperature 70–1
non-imaging systems 79–82
object emissivity 65–9, 70–1
principal wavebands 70
satellite thermal IR systems 76–9
Sea Surface Temperature (SST) 73
spatial variability 70
thermal crossover 72
thermal IR spectra 79
thermal scanners 82–3, 168
thermopile IR detectors 83–4
thermosphere 121–2, 177, 184
3D sounding 186
TIMS (Thermal IR Multispectral Scanner) 79
TIROS satellites 91, 196
TOMS (Total Ozone Mapping Spectrometer) 167, 176, 190, 190. 194
TOPEX/Poseidon project 107, 145, 207
trace gas maps, Giovanni 190–1
transmittance 48, 60, 125
transmittance t 25–6, 125
transpiration, plant 73–4, 75
TRMM (Tropical Rainfall Measuring Mission) 109, 144, 166, 189, 210
TROPOMI 81, 189
troposphere 120, 175, 177, 185–90

UAVs (Unmanned Aerial Vehicles) 1, 7, 140, 155, 168, 169, 185, 230–1
ultraviolet radiation and remote sensing 9, 14–15
  and hydrosphere 47–8
  laser sounding 188
  oil spillages 147–8
  and rocks and minerals 58
  and snow and ice 59–60
  and soil 56
  and vegetation 53
uncooled FPA (UFPA) 86
UNOSAT 9, 171
urban/built environments 18, 38, 160–1, 167–8

Van Allen Probes, NASA 180–1
vegetation 18–19, 30
  canopy geometry changes 53
  effect of solar and sensor azimuth 52
  effect of solar and sensor elevation 51–2, 74

leaf structure and NIR reflectance 50–1
and microwave radiation 112
moisture content 74
and NIR and visible radiation 46
Normalised Difference Vegetation Index 18, 19, 53–5
remote sensing applications 162–6
thermal infrared radiation 73–4, 75
and ultraviolet radiation 53
visible pigment absorption 48–50
visible radiation waves and remote sensing 14
and hydrosphere 41–2, 136–7, 140, 142
oil spillage 148
and rocks and minerals 58, 155, 156
and snow and ice 58–60, 134, 166
and soil 55–6, 160
urban areas 168
and vegetation 48, 162
volcanoes 20, 70, 78, 171, 186, 189

water *see* cryosphere; hydrosphere
wavelengths, radiation 14–17
weather prediction and systems 20, 64, 72, 91, 110, 138, 144, 150, 170, 189, 194–5, 196, 203–4, 207
weather, space 179–80
web-based satellite image sources 231–3
web cams 231
wetland mapping 139–40
WideScanSAR mode 207
Wien's (displacement) law 29, 64
WorldView-3 17, 18, 20

X-rays 14

yellow band 18